D1704309

Christoph Gengnagel · Emilia Nagy
Rainer Stark
Editors

Rethink! Prototyping

Transdisciplinary Concepts of Prototyping

Editors
Christoph Gengnagel
Institute of Architecture
and Urban Planning (IAS)
Berlin University of the Arts
Berlin
Germany

Rainer Stark
Industrial Information Technology
Technische Universität Berlin
Berlin
Germany

Emilia Nagy
Hybrid Plattform
Berlin University of the Arts
Berlin
Germany

"Rethinking Prototyping—New Hybrid Concepts for Prototyping" was a transdisciplinary research project of the Technische Universität Berlin and of the Berlin University of the Arts, funded by the Einstein Foundation Berlin. www.rethinking-prototyping.org

Translation: ZIS Fachübersetzungsservice, www.fachuebersetzungsservice.com
Copy Editing: Ralf Sonnenberg and Mark Kanak, www.lektoratberlin.net

ISBN 978-3-319-24437-2 ISBN 978-3-319-24439-6 (eBook)
DOI 10.1007/978-3-319-24439-6

Library of Congress Control Number: 2015952031

Springer Cham Heidelberg New York Dordrecht London

Printed on acid-free paper

Springer International Publishing AG Switzerland is part of Springer Science+Business Media (www.springer.com)

Foreword

When we purchased a car in the past, we selected a standardised model that we liked more or less. Henry Ford is credited with saying: "Any customer can have a car painted any colour that he wants so long as it is black". If we buy a new car today, we can choose from dozens of seat features, glazing and electronic assistance systems, and so on. Products are being adapted more and more to be compatible to our specific needs. Quite soon, we will probably design the cars we purchase ourselves. A car becomes a so-called hybrid performance bundle where the manufacturing and the service components merge together. At the same time, the danger of overwhelmed consumers who can no longer cope with the wide range of possibilities and the consequences of their choice certainly exists. In the project called "Rethinking Prototyping", designers and engineers examined the requirements that the development of hybrid products impose on them and the role that prototypes will assume in the future. "Rethinking Prototyping" is one of 30 projects that are currently funded by the Einstein Foundation Berlin (Einstein Stiftung Berlin). What were the criteria for support?

The Einstein Foundation funds outstanding scientific and creative projects in Berlin, on the highest international level, a special feature of the foundation. All the Berlin universities and the Charité—Universitätsmedizin Berlin (Charité University Medicine) are entitled to submit an application. Enabling Berlin to shine like a beacon at the summit of research and simultaneously reinforcing the creative potential of the city are the goals that the Einstein Foundation has adopted. The Einstein Foundation considers itself a partner of universities in Berlin. The subsidy programmes should act as a catalyst in collaboration with various institutions, fields and research groups, as well as provide significant added value for the city as a result. There are no quotas in terms of disciplines or institutions.

Einstein research projects are characterised by the fact that they are innovative and potentially risky in the most positive sense, while being supported by multiple institutions in the city. In each round of funding, the foundation receives significantly more good applications than can be approved. "Rethinking Prototyping" was able to prevail in a highly competitive process before renowned professional experts:

the project meets the criteria of the Einstein Foundation in an exemplary way. Interdisciplinary research teams at the Berlin University of the Arts and the Technische Universität Berlin carried out the three-year research project jointly. It was not only the first project on the "Hybrid Plattform", a joint network involving the UdK Berlin and the TU Berlin, but also the first project funded by the Einstein Foundation that brought together engineers and designers.

The underlying transdisciplinary approach defined the project. Close collaboration between various disciplines required that the participants question and overhaul their own conceptual patterns and knowledge concepts. New concepts for prototyping were supposed to result from the synthesis of knowledge from various disciplines. The results will now flow into research and teaching and be made available to the creative economy in Berlin.

This final publication presents the knowledge gained from the "Rethinking Prototyping" project and the conclusions that it permits for the transdisciplinary approach. For the Einstein Foundation, one thing is certainly clear: it was a courageous path that the applicants endeavoured to take, and it will be necessary to support these transdisciplinary approaches in the future, as well.

We hope you find the publication stimulating.

Prof. Dr. Dr. h.c. mult. Martin Grötschel
Einstein Foundation Berlin

Preface

This book is the result of three years of intensive research work on prototyping. The cooperating fields and departments shared a scientific interest in both new approaches to development processes and a collective understanding of prototyping. The project called "Rethinking Prototyping—New Hybrid Concepts for Prototyping" brought together fields that often differ greatly in terms of their methodology, terminology and hypotheses or even partially contradict each other: the engineering disciplines at Technische Universität Berlin and the artistic design disciplines at the Berlin University of the Arts.

In this project, the researchers embarked on a journey into uncharted territory since there was often a lack of established interdisciplinary experience between the fields. It was common practice to think laterally and outside the box in the three interdisciplinary sub-projects entitled "Hybrid Prototyping", "Blended Prototyping" and "Beyond Prototyping". The theoretical and practical exchange was not solely on a research level; however, all the researchers were involved in the establishment of a discourse wherein the generally formulated search for a collective understanding of prototyping played the main role. In practice, transdisciplinary research required additional reflection from all participants in regarding their own perspectives and openness in considering other scientific points of view. The process of finding suitable research formats and methods was itself the object of much consideration during the course of the project. The experiences and scientific knowledge that resulted from this intercultural experiment is documented in this book and also includes scientific contributions from guest experts whose ideas stimulated and supplemented the research as well, both directly and indirectly.

This publication itself is the result of a searching process. For the prototypical, self-reflecting and optimising overall project, we sought an appropriate format that reflects the character of the experimental research and in which the results are manifested in part plastically and thus become perceptible. We realised the solution in a comprehensive package that includes this book and other objects that were developed in the prototyping process. The analogue and digital elements of this so-called layer cake impart knowledge in a playful, appealing and generally understandable way, all stemming from extensive reflections on a common understanding of prototyping. You will find more information on the entire package in this book (Ängeslevä et al. "Results of Rethinking Prototyping") and on the project's Website www.rethinking-prototyping.org. The "layer cake" will be released in a limited edition.

One of these layers is the book that you have in your hands. It is a hybrid anthology that has high expectations for integration. It shares the transdisciplinary scientific knowledge gained in this research project that can be used for future basic research and also has potential for future application. In addition to the articles on the results from the research project itself, the anthology also provides a multi-perspective view of the broad theoretical and practice-relevant field of prototyping through inclusion of additional external perspectives provided by guest experts. Furthermore, this book also describes the experiences gained from the coordinating support for the project, which can be useful for the planning of future interdisciplinary projects.

The conclusion of a research project manifests itself, as a rule, in a scientific publication. This publication and its findings and observations, however, always remain only one static picture of an ongoing discourse that is also needs to continue regarding prototyping research. In this sense, we wish you an exciting and inspiring read and hope it encourages you to reflect and to continue rethinking prototyping.

Berlin
July 2015

Christoph Gengnagel
Emilia Nagy
Rainer Stark

Acknowledgments

In the name of all the project participants, we would like to thank everyone who supported the project in an advisory capacity and enriched it with their ideas and suggestions. We would like to thank Prof. Jörg Petruschat for sharing his thoughts and opening up new perspectives on the research question and Prof. Johann Habakuk Israel for his major contributions in shaping the project at the beginning as well as his continuous conceptual advice and active participation. We would like to thank Dr. Maria Oppen for illuminating and constructive discussions on the support and design of this transdisciplinary project. For the enriching collaboration and for the constitutive framework of the project, we also owe special thanks to the Hybrid Plattform and its coordinators Claudia Müller and Julia Warmers. We would also like to thank the Einstein Foundation Berlin for its financial support and its attentive and supportive role in the project.

Our appreciation is also extended to all the experts who took part in our workshops and were very important sources of inspiration for this project. We thank the following individuals very much for sharing their enriching and thought-provoking ideas.

Julian Adenauer/re:tune, sonice
Mathias Graf von Ballestrem, Dr. Ing./Technische Universität Berlin
Luis Berríos-Negrón/Artist
Simon Deeg and Andres Picker/studio milz
Christina Dicke, Ph.D./IXDS Berlin
Sarah Diefenbach, Prof. Dr./Ludwig-Maximilians-Universität München
Moritz Fleischmann, Prof./Henn Architekten, Berlin
Anja Götz and Tom Ruthenberg/IONDESIGN Berlin
Marguerite Joly/Ullstein Buchverlage GmbH, Berlin
André Kalms and Deon Venter/Rolls-Royce Deutschland Ltd. & Co.
Pauline Klünder/Design Report, Berlin
Kristin Kohler, Prof./Hochschule Mannheim
Tim Peix/Kunsthochschule Berlin-Weißensee
Leonard Streich/something fantastic, Berlin

Contents

Contributors

Jussi Ängeslevä Institute for Time Based Media, Berlin University of the Arts, Berlin, Germany

Benjamin Bähr Quality and Usability Lab, Technische Universität Berlin, Berlin, Germany

Boris Beckmann-Dobrev Industrial Information Technology, Technische Universität Berlin, Berlin, Germany

Mark Burry Melbourne School of Design, Faculty of Architecture, Building and Planning, University of Melbourne, Melbourne, Australia

Ulrike Eichmann Hybrid Plattform, Berlin University of the Arts, Berlin, Germany

Konrad Exner Industrial Information Technology, Technische Universität Berlin, Berlin, Germany

Christoph Gengnagel Institute of Architecture and Urban Planning (IAS), Berlin University of the Arts, Berlin, Germany

Johann Habakuk Israel Computer Science, communication and Economy, University of Applied Sciences, Berlin, Germany

Axel Kilian Princeton University School of Architecture, Princeton, USA

Kora Kimpel Institute for Time Based Media, Berlin University of the Arts, Berlin, Germany

Simon Kind Industrial Information Technology, Berlin, Germany

David Lindlbauer Computer Graphics, Technische Universität Berlin, Berlin, Germany

Kai Lindow Industrial Information Technology, Technische Universität Berlin, Berlin, Germany

Sebastian Möller Quality and Usability Lab, Technische Universität Berlin, Berlin, Germany

Emilia Nagy Hybrid Plattform, Berlin University of the Arts, Berlin, Germany

Iohanna Nicenboim Institute for Time Based Media, Berlin University of the Arts, Berlin, Germany

Rainer Stark Industrial Information Technology, Technische Universität Berlin, Berlin, Germany

André Sternitzke Institute of Architecture and Urban Planning (IAS), Berlin University of the Arts, Berlin, Germany

Martin Tamke CITA, Centre for Information Technology and Architecture, KADK, Royal Danish Academy of Fine Arts School of Architecture, Copenhagen, Denmark

Mette Ramsgaard Thomsen CITA, Centre for Information Technology and Architecture, KADK, Royal Danish Academy of Fine Arts School of Architecture, Copenhagen, Denmark

Jens Wunderling Institute for Industrial Design, Interaction Design Studies, HS Magdeburg-Stendal, Stendal, Germany

Introduction

Christoph Gengnagel, Emilia Nagy and Rainer Stark

Ideas and approaches for practical solutions become manifest in prototypes. They enable us to consider and test them as well as to communicate about them. Prototypes inspire new ideas, demonstrate problems and let us test solutions. They are tools in the creation, development and design process, which have traditionally been shaped in different ways depending on the field. How does the process of prototyping change if representatives from different prototyping cultures are involved in a project if engineers, designers, architects and software engineers work on the production of joint prototypes from the beginning? The project entitled "Rethinking Prototyping—New Hybrid Concepts for Prototyping" let the involved institutes and researchers open up a large area of reflection on their individual and joint action in the process of prototyping. The participating researchers asked why and how prototyping can be re-conceived and whether a cross-disciplinary definition of prototyping is possible, and what this would be called.

The "Rethinking Prototyping" research project was one of the first joint long-term research projects by the Technische Universität Berlin (TU Berlin) and

C. Gengnagel (✉)
Institute of Architecture and Urban Planning (IAS),
Berlin University of the Arts, Berlin, Germany
e-mail: gengnagel@udk-berlin.de

E. Nagy
Hybrid Plattform, Berlin University of the Arts, Berlin, Germany
e-mail: emilia.nagy@hybrid-plattform.org

R. Stark
Industrial Information Technology, Technische Universität Berlin,
Berlin, Germany
e-mail: rainer.stark@tu-berlin.de

C. Gengnagel et al. (eds.), *Rethink! Prototyping*,
DOI 10.1007/978-3-319-24439-6_1

the Berlin University of the Arts (UdK Berlin).[1] It was carried out on the "Hybrid Plattform", which is the joint transdisciplinary project platform of the two universities,[2] incubator and trailblazer of projects that go beyond the limits of the individual disciplines and universities. The project was undertaken in this specific framework, covering the productive area of interest between the engineering sciences at the TU Berlin and the creative-artistic disciplines at the UdK Berlin.

1 Research Principle

The work in this project was carried out with a particular constellation of disciplines in accordance with the research principle of transdisciplinarity: In the three sub-projects called "Hybrid Prototyping", "Blended Prototyping" and "Beyond Prototyping", scientists from at least one field at each university researched one hybrid prototyping approach. This level was complemented by a superordinate and all-encompassing level upon which the programmatic *rethinking* of prototyping occurred through the inclusion of all the fields involved in the project. On the one hand, it was necessary to jointly formulate a concept of prototyping that could persist in all participating disciplines. On the other, the findings in the individual sub-projects were interconnected so as to generate new ideas and knowledge from the points of connection in the projects. Last but not least, the process of the joint, transdisciplinary research was the subject of much self-reflection on the superordinate, overall project level. This proved very beneficial for revealing the potential of the transdisciplinary research principle in the best possible way: the development of new concepts and the resulting motivation to replace traditional points of view and approaches in the fields and disciplines.

The transdisciplinary collaboration took place in the form of *collisions* of different work methods and ways of thinking in the various formats such as project meetings and moderated workshops. New formats for work and discussion were developed for the design of the formats thanks to the intensive researcher participation corresponding to the current transdisciplinary collaboration. They offered all the participants the opportunity to critically reflect on their own positions and to demonstrate the potential for new innovative directions. The prerequisite for this type of collaboration, reflection and innovation was the basic idea of transdisciplinarity understood as the kind of transformative and integrative inter-disciplinarity

[1]The work on the three-year transdisciplinary project was done by academics and artists from the Design Department of the UdK Berlin in the fields of "Design with digital media", "Design research" and "Constructive drafting and architecture planning". The participants from the TU Berlin included the departments of "Traffic and machine systems" and "Electrical engineering and computer science", with the fields of "Industrial information technology", "Computer graphics" and "Quality and Usability", as well as the associated research institutions at the Fraunhofer Institute for Production Systems and Design Technology (IPK).

[2]For more information about the "Hybrid Plattform", see www.hybrid-plattform.org.

that was implemented throughout the entire project as a working and organisational principle. In terms of the transformative aspect, this project aimed at inspiring the disciplines involved by providing new ideas through the participants and thus also transforming them over the long term, e.g., with respect to the defining discourse and focal points. In terms of the integrative aspects, the project should create the possibility of developing and answering new questions that have not been clarified by disciplines hitherto—by means of a transdisciplinary integration of knowledge.

Prototyping in this project was understood as a researching and creative activity that cannot be categorised under a discipline, but rather allows and addresses diverse and overarching questions. These were developed and handled on two levels in the project: one, by inclusion of the theoretical and internal-scientific discourse of the disciplines involved (*theoretical transdisciplinarity*), and the other within the scope of the practically-oriented research in the sub-projects, the guiding questions of which arose outside the academic institutions (*practical transdisciplinarity*) (Mittelstraß 2005). Non-academic actors were also invited for their application-oriented and system-overarching answers, and thus another important aspect of transdisciplinarity was practised and researched experimentally: the mutual exchange of knowledge between science and society.

In summary, the "Rethinking Prototyping" project combined goal-oriented, practice-related development work, experimental basic research and a reflexive theoretical approach. Methodologically, it offered space for both creative-speculative ways of working and an analytical and systematic procedure, revealing new connections from both.

2 Rethinking Prototyping in Context

The culture of designing and product development is in transformation: "Traditional" products, in terms of purely functional objects, are being created more and more infrequently. New products are increasingly being linked to local or global infrastructures and integrated into them. The growing networking of all areas of life is reflected in modular and intelligent systems that complement traditional products and partially replace them. One of the most important functions in these networked or potentially networked multi-component systems is the two-sided interaction with the user, and also with other objects or systems. The primary functionality of an object is expanded by functionalities that let it acknowledge itself as a system that is a part of other systems. On the other hand, the experience and the interactive discovery by the user take centre stage. This trend can be observed, for example, in product service systems, intelligent buildings, in smart phones, in medical technology or in self-driving cars. This aspect is the focus of the sub-project entitled "Hybrid Prototyping".

The traditional term "product" is not appropriate for these systematic and interactive implications of technical innovation. This is due to the fact that these new products contain condensed competencies possessing a significantly greater scope than has been included in traditional products of the past. New product concepts require a new development culture. The traditional prototyping previously applied by various disciplines, which was defining in each case, is not capable of meeting the new demands. In order to follow this trend, i.e. to develop interactive intelligent systems, it is necessary to bundle the competencies of multiple disciplines. The focus is less on the prototype and more on the dynamic prototyping process, which allows an effective understanding to be reached between the actors, a description of a variety of solution variations in a short time and quick validation and decision-making. In the sub-project called "Blended Prototyping", a new format was developed for this very situation.

The traditional understanding of the product also gives the impression of completeness and is based on the idea that a product is the ideal end point of a development. Now the idea of an endless chain of preliminary versions (*permanent beta*) is increasingly becoming the norm in all areas of product development. The term "product" is being rendered obsolete: The product remains a prototype, and the prototype becomes a product. Rapid manufacturing methods, for example, which have been used in the generation of prototypes so far, assume an important role in the production of things that can be purchased. However, the object that can be sold in these cases is not solely the thing itself, but rather also the involvement of the future users or the end consumers in the process of development. The buyers pay for their inclusion in the sequences of the prototyping process, which is simultaneously a production process, by having their individual ideas and needs integrated into the design. The sub-project called "Beyond Prototyping" addressed this technological and social development.

In order to effectively design the development cycles in multi-competent teams, one needs hybrid prototyping concepts and the corresponding skills of the involved actors to be able to act in a transdisciplinary context. Only with close collaboration between multiple fields can the complexity of the requirements be reflected in realisation concepts. This demonstrates the need to reconsider, recombine and refine the current prototyping processes, particularly in interdisciplinary and transdisciplinary research teams.

The programmatic *rethinking* in this project aimed at preparing a general concept of prototyping and reconsidering the entire development process with respect to changes in product understanding in special individual constellations. In the sub-projects, the participants developed, tested and researched new combined prototyping concepts that can be used in specific development processes with hybrid requirements.

3 From Prototype to the Hybrid Prototyping Concept

The term "prototype" (etymologically derived from the ancient Greek word *protos*, "the first" and *typos*, "archetype or model") plays a central role in the engineering, development and design processes of all the fields included in the project.

It describes—according to the general understanding—a material or virtual object, or an experimental arrangement, simple or more complex functionality in which an idea to be realised is manifested in different stages of development—in part only in its selected properties and components. One main characteristic is that all fields also recognise its use in an iterative optimising development and work process in which the prototype fulfils different functions as a communication medium and as a model for inductive-analytical work processes.

The existing pragmatic approaches for the prototype in the fields and disciplines cannot be condensed into a discipline-overarching definition of the prototype. The respectively established field-specific understandings of the function, the area of use, the objective and the necessary complexity of the prototype in the prototyping processes differs in part significantly. Four main areas of use for prototypes and the connected functions of prototyping could be identified in this project in the summary of the involved sub-areas and from the ideas of the guest experts: (1) generating ideas, (2) user perspective and expectations, (3) communication and (4) validation and testing. The different approaches to these categories have traditionally been condensed into the difference between the *technological prototype* and the *design prototype*: The technological prototype is intended for proof on the object as well as testing and confirmation of a planned and deterministically-specified functionality. This prototype acts as a form of instructions for various serial manufacturing. At the beginning of a development project, a design prototype serves to externalise an idea, determining the target horizon and defining the problem. In later development phases, a design prototype also primarily involves functionality, however the aspect of use, interaction and communication takes precedence here. Questions about the acceptance and the needs of users as well as the complexity and sequence of actions should be answered on the basis of the prototype.

Prototypes are idea-generating tools on the one hand, and argumentative, demonstrative and analytical instruments on the other. The traditional pair of terms—the technological prototype and the design prototype—can be translated on the level of the prototyping processes as a dichotomy between analytical and generative prototyping. This clear classification no longer functions sufficiently well, however. The necessary bundling of competencies in the development processes will also allow the analytical and generative processes to be merged into a hybrid prototyping process in the future. In this hybrid prototyping process, prototypes become a medium of holistic and systemic designing.

4 Content of This Volume

Part I, "Perspectives on Prototyping", draws on the conceptual diversity of prototyping and opens up a broadly outlined perspective on the subject. It takes up both the project-related approaches and the ideas of external experts. In the first chapter, Johann Habakuk Israel, Benjamin Bähr and Konrad Exner map out differences in the understanding of the term "prototyping" and consider how prototyping could develop out of a continuous interdisciplinary exchange in the future. Of fundamental importance for its development are the central findings and experiences that were gained from scientists and practitioners in an interdisciplinary expert discussion. The second chapter by Kora Kimpel introduces various design prototyping formats that allow for a user-centred research plan and technological development, and discusses them on the basis of concrete project examples. The author emphasises two main aspects: as an interdisciplinary form of communication, prototyping from design can create a collective understanding of the use of technology and enables the aspirations and requirements of future technology to be determined. Axel Kilian responds in his chapter to the development of computational constructs spanning the physical and digital realm. This opens up a new domain the author calls embodied computation. In his article a number of prototypes are discussed in developing the concept of embodied computation through material and actuated constructs. The chapter by Mette Ramsgaard Thomsen and Martin Tamke deals with the interdependence between the digital and the physical prototype. The authors explore how the digital informs the physical and how the physical informs and interacts with the digital. Mark Burry reflects on the limits and possibilities of prototyping in architecture especially with respect to digital prototyping methods. Those allow for very innovative, but also unfamiliar space designs. By drawing on recent revelations from the continuing construction of Gaudí's Sagrada Família Basilica, the author argues that unfamiliar architectural designs cannot be properly visualized or prototyped to be experienced by the "user" before its ultimate realization.

Part II, "Rethinking Prototyping", describes the results from three years of research in the three sub-projects and from the superordinate rethinking of prototyping. Kai Lindow and André Sternitzke describe the connection between the sub-projects and the prototyping concepts represented in them. Their article shows the links found between the prototyping concepts of the fields represented in the project and describes the cooperation between the sub-projects. The results of the sub-projects are presented in individual articles. Konrad Exner, André Sternitzke, Simon Kind and Boris Beckmann-Dobrev, researchers of the sub-project "Hybrid Prototyping", present new prototyping approaches for the integrated securing of product service systems (PSS), i.e. holistically-developed systems from benefits in kind, services, infrastructures and business models. The prototyping of such complex systems is handled in accordance with the smart hybrid prototyping, a technology that makes the digital models and physical prototypes interactively perceivable in a virtual reality. In the project, these innovative prototypes were created for developments in the area of urban mobility and evaluated with customers of such PSS. Benjamin Bähr and Sebastian Möller describe the results from the "Blended Prototyping" sub-project. The project addressed new processes whereby prototypes can be designed and developed for user interfaces of mobile apps. The focus was shifted away from the computer tools usually used in this context and toward new paths of design where paper sketches play a central role. With constant feedback from app designers and developers, the approach was developed into a practically applicable process that was successfully evaluated within the framework of a user study as a creative tool for draft processes in groups. The third sub-project "Beyond Prototyping" examined the question of prototyping in connection with algorithmically generated design from the point of view of the production technology called rapid manufacturing. In this instance, the prototype becomes obsolete; end users design each product as a unique item by setting the cornerstones for the generation of the external form by themselves. Jussi Ängeslevä, Iohanna Nicenboim, Jens Wunderling and David Lindlbauer reflect on a great number of ideas and functional prototypes stemming from the university courses taught during the research project "Beyond Prototyping", as well as three case studies that present a vision for algorithmically defined products, where the dialogue between the designer, the manufacturing process and the customer can be structured differently than before. The last chapter of Part II "The Results of Rethinking Prototyping" specifies the insights regarding a possible joint understanding and a general concept of prototyping.

Part III, "Reflections on Transdisciplinary Research", considers the process of joint transdisciplinary research. Ulrike Eichmann and Emilia Nagy describe the transdisciplinary research principle, starting with its central characteristic of knowledge integration. In addition, they reflect on the process of the joint transdisciplinary research in the "Rethinking Prototyping" project from the point of the view of the coordinators by describing the major elements in the collaboration and by assessing them in terms of their potential for the promotion of the process of knowledge integration. They follow the guiding principle that each transdisciplinary project is unique and can be understood as prototypical. Each transdisciplinary project is implemented in the form of a continuous development process that is to be understood as part of a global prototyping process in transdisciplinary research.

5 Summary

As described initially, the challenges that the actors had to address in the development processes arose from a change in the understanding of the product. Classical products and services are increasingly being complemented and partially replaced by intelligent and interactive systems that are considered continuously as preliminary. In order to do justice to this change, competencies from multiple fields must be bundled. Only dynamic hybrid prototyping concepts wherein analytical and generative processes merge guarantee innovative developments. The three sub-projects in "Rethinking Prototyping" made a contribution to the hybridisation of prototyping processes in order to design interactive systems and make them possible to live out ("Hybrid Prototyping"), in order to effectively test the user experience by transforming prototypes of lower complexity quickly into higher complexity levels ("Blended Prototyping") and in order to interpret generative prototyping processes as production processes ("Beyond Prototyping"). The findings in "Rethinking Prototyping" are ground-breaking. The field-specific functions and methods of prototyping can only be clearly identified if they are able to be combined into a descriptive general definition. In the future, they will be increasingly understood as part of a holistic systematic prototyping concept. In this holistic future vision of prototyping, social-ecological components will also play a role (cf. Israel et al. in this volume).

The research in the transdisciplinary context corresponded to this hybrid character of "new" prototyping. The results of the sought integration of knowledge, the new findings in the research on the one hand, and the initiated transformation of the involved disciplines and fields on the other can be found in the articles in this volume. This project also made an important contribution to the (continuing) training of the participants' individual multi-field competencies for integrating knowledge from various disciplinary origins in order to use the new hybrid knowledge for themselves, for the joint research question and for their own discipline. As a result, "Rethinking Prototyping" made a contribution both to the hybrid and holistic prototyping and to the future of transdisciplinary research.

Reference

Mittelstraß, J. (2005). Methodische Transdisziplinarität. *ITAS Zeitschrift Technologiefolgenabschätzung. Theorie und Praxis, 14*(2), 18–23.

Part I
Perspectives on Prototyping

Perspectives on Future Prototyping—Results from an Expert Discussion

Johann Habakuk Israel, Benjamin Bähr and Konrad Exner

Abstract The role of prototyping in today's product development processes has been examined in numerous empirical studies and investigations. In various disciplines, prototyping is understood as a significant methodology for supporting clarification, conception, and design phases. Due to this significance, the question how prototyping will evolve in the future is of high relevance for those who are planning development processes, developing prototyping tools and for design researchers generally. However, quite little is known about possible future evolutions in prototyping and only few authors explicitly address this topic in the literature. This article explores perspectives on future prototyping based on the results of a focus group discussion that was conducted amongst ten prototyping experts from academia and industry. The results suggest that prototyping will maintain and even expand its general importance for product development processes. Moreover, significant changes are expected in the fields of prototyping design methods, prototyping technologies, and societal impacts of prototyping.

J.H. Israel (✉)
Computer Science, Communication and Economy, University of Applied Sciences, Berlin, Germany
e-mail: JohannHabakuk.Israel@htw-berlin.de

B. Bähr
Quality and Usability Lab, Technische Universität Berlin, Berlin, Germany
e-mail: baehr@cs.tu-berlin.de

K. Exner
Industrial Information Technology, Technische Universität Berlin, Berlin, Germany
e-mail: konrad.exner@tu-berlin.de

C. Gengnagel et al. (eds.), *Rethink! Prototyping*,
DOI 10.1007/978-3-319-24439-6_2

1 Introduction: Prototyping Definitions, Processes and Tools

Prototyping is used in various disciplines in academia and industry. Even in everyday life and popular science prototypes are commonly referred to. Nevertheless, the meaning and connotation differ widely, thus describing various characteristics. In many disciplines the term prototyping carries its own meaning and connotation. In product development, prototyping is a means to assure particular product features, e.g. stability or ergonomic functions (Stark et al. 2009), here the prototype should be as precise as possible. During the design process the prototype evolves to become the final product (Kamrani and Nasr 2010). Additionally, the digitization of the engineering design process facilitates the increased application of digital prototypes (Adenauer 2012). Prototyping in human computer interaction describes the process of creating interfaces variants and access their characteristics and qualities while developing the user interface (Pering 2002). The prototype should both be interactive and flexible in order to develop variants on the fly (Buxton 2007). The evaluation of design failures with the customer is the main objective (Lim et al. 2012). In architecture, prototype and final product are basically the same. Because the costs for building a physical one-to-one prototype of a building are high, prototypes that aim at assessing the most important usage features are basically not affordable. However it is possible to asses and verify certain details of the construction, i.e. bent structures, fittings, and material (Gengnagel et al. 2013).

Due to the various perspectives, a common cross-disciplinary definition of prototyping or prototyping processes was not developed yet. Instead, generic descriptions and conceptual frameworks exist. An example by Warfel (2009) which emphasises the duality of prototyping and validating is presented in Fig. 1.

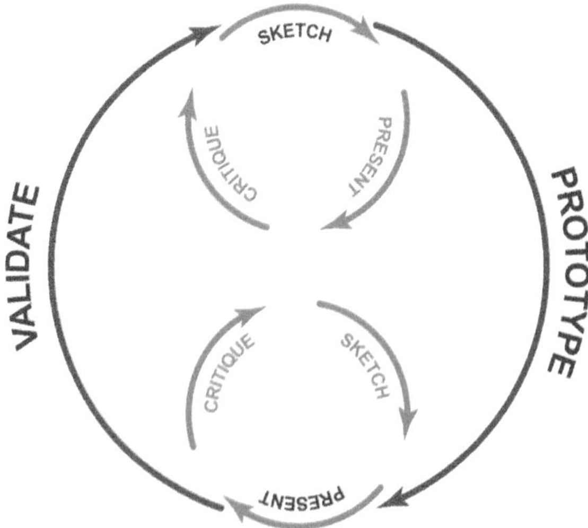

Fig. 1 Prototyping process (Warfel 2009)

Reflecting prototyping from a generic perspective enables discussions between disciplines to occur, but might simplify prototyping in an over-exceedingly manner. Therefore, in addition to the exploitation regarding the discipline specific perspectives on prototyping, further characteristics need to be considered. In an effort to develop a transdisciplinary perspective onto the prototyping process, Exner et al. (2015) conducted a case study among prototyping experts from different disciplines. They investigated multiple dimensions of the prototyping process regarding specific perspectives:

- Objectives (explorative, experimental, evolutionary),
- Dimensions (form, material, concept, principle, process, functions, requirements) and
- Fidelity (high, low, and mixed).

The process model developed by the authors integrates different perspectives from the disciplines, thus providing a basis for a common understanding and encouraging communication amongst design researchers (see Fig. 2), even though neither specific methods nor tools were reflected in the study.

To assess the tools and processes design teams are currently using, McCann (2015) and colleagues conducted a survey amongst 33 designers of interactive products from popular brands, mainly web-based and mobile products. They found that for creating prototype's contents (assets), 50 % of designers are using only one tool, most prominently a 2D digital sketching tool. In order to add interactivity into the prototypes, digital 2D-based frameworks are used to implement high fidelity prototypes, whereas many designers still use standard office tools for low fidelity prototypes. The surprisingly small diversity of tools and the partly insufficient technology suggests that the integration of prototyping technologies and methods

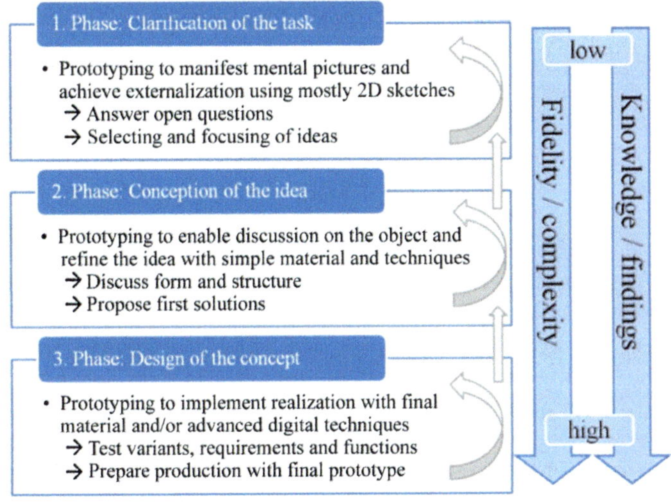

Fig. 2 Prototyping process (Exner et al. 2015)

into the process chain is still underdeveloped. Whether this can be ascribed to technological, conceptual or other reasons is an open research topic that needs to be analysed in future.

2 The Future of Prototyping in the Literature

Only few authors explicitly address the future of prototyping in design and development. Aycan and Lorenzoni (2014) propose live prototyping as a future approach in addition to existing approaches that from their perspective include rapid prototyping, technical prototyping and pilots. Live prototyping "involves releasing still-rough concepts into the context where consumers would eventually encounter them during the course of their daily routines" (Aycan and Lorenzoni 2014). Be it a store shelf or an app store, the prototype has to be encountered by the consumer between all competing choices and distractions. The natural behaviour of the consumer is observed before intercepts and interviews are conducted. The authors suggest that live prototyping conserves capital relative to a full pilot, considers the context, improves forecasting and provides qualitative and quantitative feedback. Aycan and Lorenzoni point out that applying live prototyping has to take cultural norms into account. "While American consumers have shown a hunger to co-create solutions with companies and tend to celebrate brands that embrace experimentation and that are 'permanent beta', this is not always true in global markets. It's important to calibrate what degree of 'roughness' is going to be acceptable based on the market in which you're operating" (Aycan and Lorenzoni 2014).

Some authors, e.g. Hodges et al., emphasise the role of technical prototyping platforms in the future (Hodges et al. 2013). It is supposed that such platforms will speed up the development of prototypes, support the transition between prototypes of various maturity levels and different materials, as well as contribute to the variety of prototyping tools and components (Hodges et al. 2013).

Looking at the future of prototyping, Schrage (2000) stresses the importance of shared spaces for the development of new insights about product ideas and organisations. He suggests that collaborative methods such as serious play will achieve a high share among future prototyping methods.

Blomkvist et al. (2011) emphasise that prototyping has been accepted as a holistic design technique today, but that particular deficits need to be addressed and new prototyping techniques and approaches need to be developed. Based on a literature review, they identified the most critical points with respect to prototyping in the fields of user experiences, contexts, and social interactions. They argue that especially the validity (i.e. the degree of similarity in test and implementation contexts) and the author (i.e. the important perspective of the prototype's author and the user and customer possibilities for participation in the prototypes' creation) need to be investigated in further research. Kora Kimpel introduces another perspective in her contribution "Design Prototyping for planning research and technology development" in this volume. She suggests employing three classes of prototyping with differing degrees of determination and openness regarding the applications fields and technologies used, namely design prototyping, co-prototyping and participatory prototyping.

In summary, the authors who investigated future directions of prototyping mainly addressed process issues (e.g. live prototyping, serious play), whereas on the technical side, frameworks were suggested for fostering the ease of prototype development. However, the empirical foundations of such works need to be strengthened in order to stimulate the development of new tools, technologies and process models for future prototyping.

3 Focus Group Expert Discussion

Taking the phrase *rethinking prototyping* literally, a qualitative study was performed with the objective to investigate possible roles and technologies of prototyping in the future. The aim of the study was to broaden the view onto this topic and to include external perspectives from academic and industrial experts. Even though a comprehensive depiction of the theme was beyond the scope of this study, the study was set up to discuss the future of prototyping in general, without limitations to particular domains or application areas.

In order to approach the aims of the study, the focus group method was chosen. Focus groups are performed in interactive group sessions among persons from specific target groups. The sessions are led by experienced moderators who ensure the progress of the discussions, but do not introduce their own opinions or positions. Focus groups are efficient methods for qualitative research. They are well established and accepted for gaining insights and information that would be less accessible with other, less interactive methods (Krueger and Casey 2000).

Prior to the focus group expert discussions a semi-structured guideline was developed by involving representatives from all sub-projects of the "Rethinking Prototyping" project. The guideline included open research questions to be addressed in the discussion. All questions were discussed in the preparation team. Critical questions were simulated in mock-up discussions. Finally, only those questions that passed the plausibility check were included in the questionnaire.

Furthermore, stimulus material taken from preliminary results of the sub-projects "Hybrid Prototyping", "Blended Prototyping" and "Beyond Prototyping" (Rethinking-Prototyping 2015) was prepared in form of short presentations, which included pictures and video material. The presentations were held by the moderators and co-moderators to provide impulses to the discussions.

3.1 Subjects

The focus group interview was conducted with ten experts who were selected from the fields of industrial engineering (2 participants), interaction and service design (1 participant), product design (2 participants), and academia with strong records in prototyping research (5 participants). Participants received no compensation for their expenses. Participants included authors of books and conference papers about prototyping, leaders of large third-party funded scientific projects about prototyping and design engineers from globally operating manufacturing companies.

3.2 Procedure

The focus group session was held and protocolled in German. One moderator and two co-moderators led it. The moderator was responsible for the progress of the session. The co-moderator kept an overview and ensured that all topics from the guideline were covered. The second co-moderator protocolled key statements, functions and visions (see Sect. 3.4.2) on flip charts in form of mind maps. The moderator had little influence on the content of the discussion but intervened whenever it was close to losing focus or veering off topic. The session lasted four hours, including a break after two hours. It was videotaped and audio recorded; one co-moderator took a handwritten protocol.

After a short introduction of the moderators and a brief introduction into the aims of the study, participants introduced themselves and explained the role prototyping is playing in their daily life. Each participant had enough time to introduce her or his individual perspectives and experiences. Afterwards, questions related to the topics "functions of prototyping" and "the process of prototyping" were discussed. The stimulus material was then shown and opened up the discussion about "visions", i.e. "new technical possibilities for prototyping" and "the future of prototyping". At the end of the focus group sessions, participants were asked to substantiate their ideas about "future prototyping methods and practices" on cards that were clustered on pin boards.

3.3 Analysis

After the focus group, the written protocols, the content of the flip charts and the cards written by the participants were carefully analysed, aggregated, structured and interpreted and finally discussed among the moderators and another project member in order to form a common perception of the content and answers to the main research questions (Mayring 2003). Thus, the results reflect both the ideas developed during the verbal discussion and the ideas written on the cards.

3.4 *Results*

3.4.1 Statements

Among the industrial representatives there was no doubt that prototyping is of outstanding importance for their personal work and the development processes in their businesses. All academic representatives shared this opinion and referred to corresponding research results. Some statements included: "We use prototyping from little foam models to large milled or printed products. I find it exciting to think about the diversity of possibilities in it."; "Among all the design activities which we perform, prototyping is a tool which we use every day."; "During the design process with customers we use prototypes to retrieve the current status; this is essential."

Different perspectives regarding prototyping became apparent when the participants described their daily practices. Those differences emerge for example with respect to costs, number and purposes of prototypes: "Prototyping is essential in our company, but it is always stands in tension with the cost-benefit relationship."; "For prototyping we use CAD software, but literally speaking all of our first engines are prototypes as they are produced in small series."; "We distinguish between prototypes which establish a space for ideation and those which can be used to evaluate something or to formulate a particular question."

Shortly after the introduction of the participants, the discussion began departing from economic, tangible advantages of prototyping (i.e. improving productivity, limiting failures etc.) toward the benefits of prototyping for personal and societal development, as well as the dangers of prototyping in supporting the economics of growth: "In the past we built our prototypes with foam. Today we can model them using 3D CAD systems and save a lot of waste."; "The world has gone haywire! What are the aims behind prototyping? Is it technical efficiency? Efficiency causes boredom!".

3.4.2 Categorisation

After categorising the participants' statements regarding the functions of prototyping today, we established five main categories: design and development, external communication, integration of the user, internal communication, and testing and validation. The function categories are listed in Table 1.

The results of the analysis of visions for future prototyping led to three main categories with respect to design methods, technology and society. The visions are listed in Table 2. The categories of both tables are different because the user statements regarding prototyping functions and visions were separately analysed and clustered. The category names are results of the clustering process.

Table 1 List of prototyping functions in today's practice

Category	Function
Design and development	– Clarify questions – Exploration – Materiality – Prototype as abstraction and/or simplification – Reduction of development costs – Reduction of development time – Sharpening the idea and/or the mental model – Variant development by means of virtual prototypes – Visualisation of concepts – Visualisation of ideas – Visualisation of the essentiality and/or the "message" of an object
External communication	– Convincing the marketing, management, and customers – Demonstration, presentation
Integration of the user	– Haptic experience – User experience – Validation of aesthetics
Internal communication	– Competence exchange – Cross-department communication through prototypes – The prototype visualises the internal structure of an enterprise – Integration of competencies of multiple persons or departments – Nonverbal communication through the prototype as physical object
Testing and validation	– Functionality testing – Increasing of the technical efficiency – Proof of effectiveness – Proof of completeness – Robustness/dysfunctions – Validation of requirements

3.5 Discussion

The prototyping functions in practice today (Table 1) as named by the participants are well covered in the literature (cf. Adenauer 2012; Exner et al. 2015; Kohler et al. 2014). They contain no surprising categories or functions. However, the list is comprehensive and emphasises expertise of the participants and their familiarity with the respect to prototyping. Given this, the following list of visions of the future of prototyping (Table 2) can be regarded as substantial.

The visions of the future of prototyping as expressed by the participants have a different and much broader scope than today's function. This can be either due to the fact that we asked the participants closed, fixed questions to describe their daily practices and open questions to express ideas and visions. Furthermore asking for input regarding visions and future-related aspects might have stimulated the participants to think in larger contexts.

A number of trends can be derived from the list of visions of the future of prototyping (see Table 2). First of all, the activity of prototyping will remain an

Table 2 Visions of the future of prototyping

Category	Visions
Design methods	1. Even greater use of prototypes as communication media 2. One-to-one functional representation of complex products and systems (a) Deep cross-module integration of prototyping sub-functions and sub-systems (b) Massive increase of virtual prototypes (c) Prototypes can be used to guarantee the functionality, reliability and informational value of future products 3. Production technologies for products and prototypes are moving closer together (a) Prototypes can go into production by means of prototyping technologies 4. Prototyping of tools (as distinguished from products) will increase 5. The environment will become the laboratory, i.e. prototypes will leave the laboratories location-based services and functions will be developed in situ
Technology	6. Generative prototyping 7. Hybrid prototyping (a) Tools which allow to combine digital and physical prototype elements in order to address all human senses 8. Materiality (a) The materiality of prototypes (i.e. their surface) become modifiable, e.g. from wood to metal to plastics 9. Quick changes between physical and virtual prototypes (a) Testing of physical interaction properties (b) Usage of new and fast rapid prototyping technologies 10. Simulation of human-prototype interaction (a) Possibilities of entirely digital prototyping without the loss of user experience (UX)
Society	11. Critical Design (a) Invocation of societal debates 12. Crowd Prototyping (a) Deployment of not-yet-finished products (beta releases) (b) The unpredictability of the users will become a driver for creative design changes of the product (c) Users/the network integrate it in their daily working and living structures 13. Modular prototyping (a) Prototyping using tested and validated sub-modules 14. New application domains (a) Printing food (b) Printing human organs 15. Open source 16. Prototypes as final products (a) Beyond prototyping 17. Prototyping for fun (the "Lego" principle)

indispensable element of the product development process. Its status is even likely to increase (1) and reach the customer (12, 17). None of the current prototyping functions were explicitly designated or earmarked to become superfluous in the future. However, no new functionalities related to creativity and the ideation process were mentioned, either. This suggests that the general prototyping process (see, Fig. 1) will not change; however a quantitative change is likely to happen, e.g. in terms of increased usage frequencies.

Furthermore, analogue and low-tech prototyping techniques, i.e. paper prototyping, were not mentioned. Today such techniques are popular because of their easiness, rapid availability, low costs etc. The fact that they were not mentioned suggests that users expect the high-tech prototyping of tomorrow (2, 8, 13) to be as easily available as today's low-tech prototypes.

Specific future technologies, i.e. holographic displays or particular 3D printing techniques, were not specified. However, new application possibilities and prototype features were mentioned, i.e. printing food (14) and modifiable material properties (8), which require new technologies. The fact that the technical realisation of such new possibilities was not mentioned suggests a faith in technology, i.e. that the participants were confident about the general technical progress and that they have a great degree of trust in the developers of prototyping technologies.

Virtuality and virtual prototypes are regarded as central building blocks in future prototyping (2b, 9). Nevertheless, virtuality alone seems not to be the sole solution, as the physical contact with the prototype was regarded as indispensable (7, 8, 9).

The societal impact of prototyping was among the most prominent topics discussed during the study. The participants see prototyping as a means to enable users (citizens) to develop products according to their (and not to the markets) needs (11, 12c, 15). On the other hand methods that involve the user in the value chain were also mentioned and partly critically assessed (12a, 12b).

The convergence of prototypes and products and the "permanent beta" attitude were also addressed by the participants with respect to production technologies (3) and deployment (12, 16). This is possibly the most radical change which can be derived from the study, as it opens up the questions of many industrial product development processes and practices for serious discussion, e.g. milestones, release and start of production dates, marketing strategies and even product lifecycle concepts, as well as personal design approaches, heuristics and strategies.

The fact that many functions of today's prototyping (Table 1) are not listed among the visions of future prototyping (Table 2) should not be read to mean that the participants think that today's functions will play no role in the future. None of the participants provide any comments in this vein; on the contrary, the relevance of prototyping as a central means for development processes in the future was unanimously emphasised.

4 Conclusions

This study was conducted as an attempt at assessing a view of the future of prototyping. The results suggest that changes can be expected in the categories of design methods, technology, and society. The participants of the study showed particular interest and expectations in future design methods. However, they were less concerned about their future technical implementation, which were considered as more or less given. Furthermore, the societal implications of future prototyping methods and techniques were actively discussed amongst the participants, particularly as they expect the importance and dissemination of prototyping techniques to spread in the future. The dissemination of prototyping in the everyday life of a society can be considered one of the most relevant changes to be expected in the future of prototyping.

Further studies are required to investigate the identified trends in depth, and to reliably predict their societal implications.

Acknowledgments We would like to thank Christina Dicke (IXDS Berlin), Sarah Diefenbach (LMU München), Anja Götz (IONDESIGN Berlin), André Kalms (Rolls Royce), Kirstin Kohler (Hochschule Mannheim), Jörg Petruschat (Kunsthochschule Berlin), Tom Ruthenberg (IONDESIGN Berlin), and Deon Venter (Rolls Royce) for their willingness to take part in the study and for their valuable input.

References

Adenauer, J. (2012). Digital tools in product development Digitale Werkzeuge in der Produktentwicklung. In Adenauer, J. & Petruschat, J. (Eds.), *Prototype! physical, virtual, hybrid, smart: tackling new challenges in design and engineering*, (1st ed.) form + zweck (pp. 216–239). Berlin.

Aycan, D., & Lorenzoni, P. (2014). *The future of prototyping is now live.* Retrieved April 12, 2015. https://hbr.org/2014/03/the-future-of-prototyping-is-now-live/

Blomkvist, J., Segelström, F., & Holmlid, S. (2011). Investigating prototyping practices of service designers from a service logic perspective. In *Nordic academy of management conference,* (Stockholm, 2011) (pp. 20–24).

Buxton, W. (2007). *Sketching user experiences: Getting the design right the right design* (in press). San Francisco: Morgan Kaufmann Publishers.

Exner, K., Ängleslevä, J., Bähr, B., Nagy, E., Lindow, K., & Stark, R. (2015). A transdisciplinary perspective on prototyping. In *International conference on engineering, technology and innovation* (Belfast, Ireland 2015). Institute of Electrical and Electronics Engineers (in press).

Gengnagel, C., Hernández, E. L., & Bäumer, R. (2013). Natural-fibre-reinforced plastics in actively bent structures. In *Proceedings of the ICE–construction materials* (Vol. 166(6), pp. 365–377). http://doi.org/10.1680/coma.12.00026

Kohler, K., & Hochreuter, T. (2014). Let's compare prototypes for tangible systems: but how and why?. In *Proceedings of the 8th Nordic Conference on Human-Computer Interaction: Fun, Fast, Foundational (NordiCHI '14)* (pp. 323–332). New York: ACM Press. http://doi.org/10.1145/2639189.2639229

Hodges, S., Taylor, S., Villar, N., Scott, J., & Helmes, J. Exploring physical prototyping techniques for functional devices using NET gadgeteer. In *Proceedings of the 7th international conference on tangible, embedded and embodied interaction—TEI '13* (New York, 2013), (p. 271). New York: ACM Press. http://doi.org/10.1145/2460625.2460670

Kamrani, A. K., & Nasr, E. (2010). *Engineering design and rapid prototyping*. New York: Springer.

Krueger, R. A., & Casey, M. A. (2000). *Focus groups—A practical guide for applied research* (3rd ed.). Thousand Oaks: Sage Publications.

Lim, Y. -K., Stolterman, E., & Tenenberg, J. (2012). The anatomy of prototypes: Prototypes as filters, prototypes as manifestations of design ideas. In Adenauer, J. & Petruschat, J. (Eds.),

Prototype! physical, virtual, hybrid, smart: Tackling new challenges in design and engineering, (1st ed.) form + zweck, (pp. 100–122). Berlin.
Mayring, P. (2003). *Qualitative inhalts analyse. Grundlagen und Techniken* (8th ed.). Weinheim: Deutscher Studien Verlag.
McCann, C. (2015). *Prototyping tools and process.* Retrieved April 13, 2015 from https://medium.com/greylock-perspectives/prototyping-tools-and-process-ab63831f8486
Pering, C. (2002). Interaction design prototyping of communicator devices: Towards meeting the hardware-software challenge. *Interactions, 9*(6), 36–46. http://doi.org/10.1145/581951.581952
Rethinking-Prototyping. (2015). Ein transdisziplinäres Forschungsprojekt der Technischen Universität Berlin und der Universität der Künste Berlin. Retrieved May 10, 2015 from http://rethinking-prototyping.org
Schrage, M. (2000). SERIOUS PLAY: The Future of Prototyping and Prototyping the Future. *Design Management Journal (Former Series), 11*(3), 50–57. http://doi.org/10.1111/j.1948-7169.2000.tb00030.x
Stark, R., Beckmann-Dobrev, B., Schulze, E. -E., Adenauer, J., & Israel, J. H. (2009). Smart hybrid prototyping zur multimodalen Erlebbarkeit virtueller Prototypen innerhalb der Produktentstehung. In *8. Berliner Werkstatt Mensch-Maschine-Systeme BWMMS'09: Der Mensch im Mittelpunkt technischer Systeme*, (pp. 437–443). Berlin: VDI-Verlag.
Warfel, T. Z. (2009). *Prototyping: A practioner's guide*. New York: Rosenfeld Media.

Design Prototyping for Research Planning and Technological Development

Kora Kimpel

Abstract Research planning and technological development are part of our ongoing social and cultural development that can be shaped in a user-centred way with prototyping models from design. As an interdisciplinary form of communication, prototyping from design can create a collective understanding of the use of technology and enables the aspirations and requirements of future technology to be determined. This added value for research planning and the development of technology is demonstrated through examples provided in this text. Various prototyping models such as design prototyping, co-prototyping and participatory prototyping are outlined as important indicators for research planning and technological development and are described in terms of their effectiveness. The respective prototyping model determines on the one hand how daily life experts can be integrated into the development process, and on the other it specifies how concretely the given model can be applied to the technology in development. Accordingly, the appropriate prototyping model must be selected for the specific issue in technological development. The detailed description of the parameters and qualities of the prototyping models as well as the graphs and visuals of them should help with these decisions.

1 Introduction

Research and technological development assume a place of particular importance in our society. This is not only the case from an economic point of view, but also because the discoveries and results from both help us to find answers to questions about the future, questions about medical care, future mobility, sustainability and the use of our resources, in short, how we want to live with each other on our planet in the future. Research can provide answers to many of these questions that can be implemented in new technologies for applications. But in all likelihood it is just as probable that

K. Kimpel (✉)
Institute for Time Based Media,
Berlin University of the Arts, Berlin, Germany
e-mail: kkimpel@udk-berlin.de

C. Gengnagel et al. (eds.), *Rethink! Prototyping*,
DOI 10.1007/978-3-319-24439-6_3

concrete proposals from the applied technological developments will not be adopted or accepted by society. There are many reasons for this. In various studies, it has been made clear that the inclusion of people in the implementation process is an important factor. In recent years, various communication models have been developed with different approaches for managing dialogues with society. Design is one discipline that focuses precisely on introducing technology into applications from a user's perspective. Design offers additional methods and procedures for this. One possibility with great potential as a communication and interaction platform is prototyping. Over the last five years, together with the Fraunhofer Center for Responsible Research and Innovation, I have worked on testing and analysing various models of design prototyping in projects for technological development. The discoveries in these processes form the basis for this text and raise the question as to how one can facilitate a user-centred research planning and technological development.

In the research project called "Rethinking Prototyping" where I am part of the "Hybrid prototyping" team, it has become clear that the prototype and the prototyping itself represent a unique possibility for negotiating processes with various disciplines and users. On the basis of prototyping, it was also possible in this project to determine and describe various research approaches in design, architecture, mathematics, computer science and mechanical engineering with their various conceptual models, procedures and decision-making criteria. Prototyping is the word of the hour. After years of intangible and digital concerns, the object itself is once again becoming the centre of attention with the Internet of Things and the new production technologies. Although the prototype is regarded less as an object and more as a process, i.e. prototyping, doing. In the process of prototyping, much has changed through the new possibilities in producing and publishing as well as the new forms of work. Processes in prototyping are more open, collaborative, discursive today, and they will become more accessible and transparent for various disciplines and thus also implementable for social discourse. In this text, the process of prototyping is considered from the design perspective, and three different models of prototyping show how prototyping can integrate the perspective of users into the research plan and technological development. The three models are introduced on the basis of examples from various research projects and are compared in terms of their potential influence for research planning and technological development.

2 Design Prototyping

The term[1] prototype is used by various disciplines to describe a certain state of a project. In design and in engineering disciplines, the term prototype is of central importance and has a comparable, but not identical purpose in both disciplines.

[1]The term consists of the words "protos", from the Greek word for first, and "typos", which means archetype or model in Greek. The prototype is often the first model in series production, but also stands for a concept draft on the basis of which you can check use and acceptance.

While engineers provide proof that a design can be achieved by producing technical prototypes, designers test the future use in the form of design prototypes. The implementation of prototypes in design is characterised by various criteria: The communication and interaction with the interface of the object to be designed and its services are in focus. Aspects of ergonomics, use and experience with the object and service, the feel, form and aesthetic are determined in the design prototypes. This prototype is an object that is being examined with regard to its use.

Engineers and designers can inform themselves about the current status of their respective discipline on the basis of prototypes and learn about important aspects in each case. What does the actual use devised in design prototypes mean for the technology and what does the used technology mean for possible new uses? A dual examination is possible with the prototype since it is more understandable for both sides, a common denominator. The prototype shows options for action irrespective of whether they are from the perspective of users or from a technological perspective. The prototype from design introduces the important parameters of the users' perspective. The design represents the connection between use, technology and aesthetic concerns, whereby aspects of economics, ecology, ergonomics and social developments and expectations are also included. Design acts on the one hand like a discipline by taking knowledge and creative competency from the design, and on the other, in an interdisciplinary way, as a negotiator and translator of the claims and tasks from society and from each of its individual representatives. Prototypes developed by daily life experts, i.e. the future user groups, require the translation of the design just as engineers must translate new discoveries in natural science research into applications so people can understand how they can be used. The focus of design is on the needs and wishes of people and society. Design gives shape to these needs. The more daily life experts are capable of visualising the needs and wishes of people and society on their own and illustrating them in the form of prototypical use scenarios and objects, the more precise the translation to the design will be. Prototyping negotiates not only between the disciplines, but also between society, science and research.

Joint Design Prototyping with Users Prototyping can be a dialogue between a person and an object, between two people or two disciplines—it is, however, also suited for dialogue within groups. A reservation voiced with regard to the joint development of technology with users and daily life experts is the opinion that users can only develop or think of things that are very close to the status quo. Henry Ford is often quoted when someone defends this thesis: "If I had asked my customers what they wanted, they would have said a faster horse"[2]. There is something true in

[2]Henry Ford (1863–1947), founder of the car manufacturer Ford Motor Company, perfected assembly line work in the automotive industry and revolutionised industrial production. He published a lot, partially in books and newspapers, gave numerous interviews and wrote many very controversial anti-Semitic texts. The familiar and polarising quote about customers and the horse has not been proven to this very day. Research by various authors has not discovered any reliable source for this quote. Since it is often cited, the quote has taken on a life of its own in the common vernacular in the meantime, which is why I use it as a conceptual model in this text.

this statement. In our daily lives, we are shaped by the reality surrounding us and cannot simply get over this. The steady flow of information, either through advertisements or reports from research and development laboratories, makes us aware of new products. The well-known refrigerator that goes shopping on its own is an idea that has often been popularised in the media in order to emphasise the added value of the internet of things (IoT) more demonstrably. If you ask users today how the internet will change their daily life in the future, you frequently hear the aforementioned scenario of the autonomously shopping refrigerator offered as an answer, usually accompanied by a shake of the head because the logic of this futuristic idea is doubted. The media has influence on what we should think about our future and makes it difficult to engage in open discourse with users on future wishes and needs. Very few people think explicitly about the more distant future, i.e. beyond their personal concerns. How should they make valuable contributions to the future if they can only think about the existing situation for a maximum of 5 years in advance—something they rarely find a reason to do—and even then, when they are already influenced by the ideas flowing out of research institutions, companies and the media? The example of the refrigerator shows, however, how important the phrasing of the question is: Do we even inquire about the future of the refrigerator or how we should handle cooling or heating in the future, period? By formulating the question in a different, more open and more active way, we gain ideas that do not replicate the status quo. The autonomously-shopping refrigerator is, in some senses, quite similar to the concept of Ford's faster horse. It is not an innovative product and certainly not a new market. The market replicates the status quo and only thinks "faster", or automatically.

While we may accuse those experts of daily life of solely operating in preconceived categories and being incapable of new ideas, we must admit that experts, designers and engineers move in defined fields and formats, too. They do this with greater knowledge of the current developments and a larger repertoire for the development of new combinations and variations of existing ones. These are most important qualities, since the practical knowledge and the adoption of the repertoire require extensive studying and time for independent observations and discoveries, as well as the practical ability to develop something new, such as turning research questions into new discoveries in the respective disciplines. Engineers and designers, however, are also influenced by existing ideas and current discourse, for example, from science fiction and other research and development laboratories. At their core, each new idea has components that are familiar; each designer and developer recombines things, varies and scales them, transfers them to other contexts or uses and can evaluate these ideas in a broader context. The results from basic research, such as more precise measurement methods or smaller mechanics and electronics, offer designers and engineers new options that are not usually available to daily life experts. Therefore, the inclusion of users in development processes requires a method-based procedure that makes it possible to partially and temporarily adopt these specific expert competencies.

Let's return to the idea of the faster horse. It inherently possesses two interesting components that should be examined in more detail with respect to the joint development of technology with daily life experts. If we ask people about their desire for future travel, it is possible for them to consider the *means* (with what type of vehicle) or the question of how. It is possible that the *how*, i.e. the "faster" is a more valuable indicator than the *means*, i.e. the horse. The fact that we can implement the "faster" better with a car than with a horse falls within the decision-making expertise of developers. For researchers and developers, the "weak" component of *how* in the form of the describing adjective is more useful than a "factual" component, such as the word horse. When user groups express themselves orally or in writing regarding the wishes and needs they have for the future, words often prove a limitation, particularly in German-speaking regions. In German, the nominal style[3] is often used in texts—meaning the language is less action-oriented or descriptive, and remains encased in familiar terminology because there is no expression for what does not yet exist. If we made it possible for users to describe their future uses on the basis of prototypical environments, we would gain access to a non-verbal world that could describe more clearly and diversely what a specific use might look like. Factors such as material qualities, size and handling would then be defined since they frequently don't find any expression on the verbal level. The "faster" is just such a factor in the case of our Ford horse. Consequently, prototyping allows us to discover ways of avoiding the "horse" and to determine the "faster". It is precisely on the haptic, tactile level that most people are less influenced by preconceived ideas. The process of shaping is accomplished in a steadier, less frenetic and more open manner than in the process of speaking. The process itself is more one of searching and also more descriptive for others. In shaping, people must indirectly answer questions that they have never been asked before. All these are factors that make prototyping with daily life experts expedient and interesting for technological development. Particularly in the areas of aesthetics and shaping, but also when it comes to a description of uses, wishes and needs, people often lack the suitable vocabulary of expression. The vocabulary employed often reflects descriptions from commercials and is limited in terms of aesthetic descriptions. That is explainable because it is unusual for most people to exchange formal and aesthetic criteria or develop their own criteria. Even in public discourse, exchanges of opinion on product design, architecture or fashion are relatively rare. It is possible to attempt to avoid these limitations in language. The action in prototyping represents a possibility of asking for the required information on a non-verbal level that would not have attracted any attention without three-dimensional representation. Interestingly, words are created in the process of prototyping and descriptions are discovered that would not have had any triggering effect without this physical interaction.

[3]The nominal style is a form of expression where nouns (called *Nomen*, or *Substantive* in German) take a larger role than verbs. This style often defines texts re-quiring a precise means of expression, such as scientific and legal texts.

3 Inclusion of User-Centred Design Prototyping in Research Planning and Technological Development

How can one speak about possible uses when one does not even know the technology? Is it not foolhardy to believe that future users' wishes and needs in technological development can be determined ahead of time? Has the history of technology not demonstrated that people will use what technology provides them only when it is present? These questions reflect a popular view in the research landscape that gives technology a head start well before a possible user comes into play. Whether the early integration of user interests can positively influence technology is a matter of debate. If one adheres to the argumentation made in the text, however, formats of prototyping can facilitate this early integration. For prototyping, this entails the following questions:

- How can technology be juxtaposed to possible uses in an early stage of development based on prototyping so that the aspects of use can be taken into account in the development?
- What formats of prototyping are conceivable for this and what benefits do they bring?

In the following, I will introduce three prototyping models[4] derived from design that demonstrate different formats for the inclusion of user perspectives in technological development. The three models differ in terms of their methods and procedures as well as their results in regard to their specific applicability to the technology that must be developed and with respect to the amount of time and content for the inclusion of social requirements and needs:

Design prototyping can have a direct effect on development with regard to a specific field of technology. The parameters for user requirements are specific. The user-specific requirements and needs are included and evaluated by studies and method-based observations.

Co-prototyping should be implemented at an early stage of technological development since it can have a significant influence on the development. User-centred and social requirements and needs are queried by means of co-design processes and evaluated so that specific parameters can be described for the technology, but previously unconsidered fields of development and markets become visible.

[4]The three models were developed in the context of research projects that were worked on with the Fraunhofer Center for Responsible Research and Innovation (Fraunhofer CeRRI) over the last four years. In the three examples, I have been involved as a project partner and accompanied the development of the methodology, the process design and the analysis in the transformation phase. The cases are exemplary for each prototyping model prior to a technological development. For each of the three formats, there are numerous other project examples from research projects at the Fraunhofer CeRRI which cannot be discussed here in full.

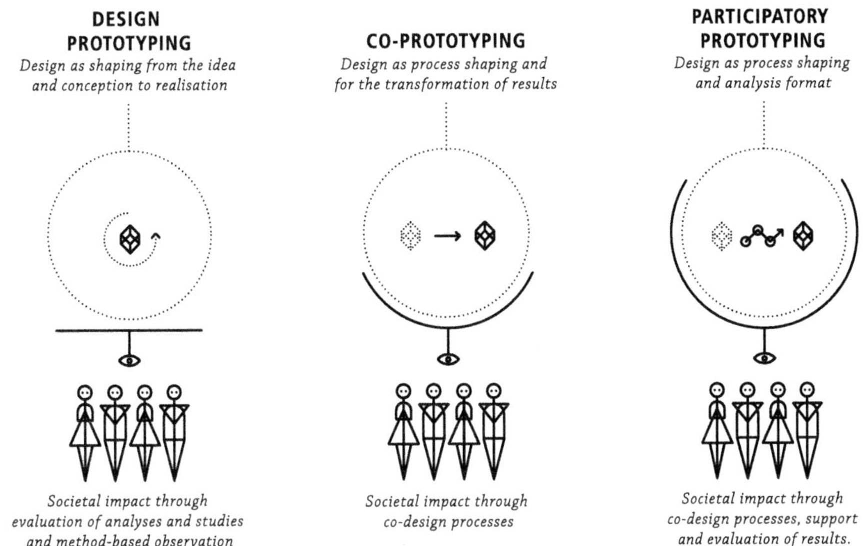

Fig. 1 Models for including a societal impact in design prototyping

Participatory prototyping should be completed prior to the development of the technology and in the research plan. It reveals potential possibilities for new fields of technology and research. Requirements and needs through usage and by society are included through processes of co-design and the supporting evaluation of development.

Figure 1 illustrates the share of the social impact over time in a typical design process. In the decisions, in the design process, socially relevant data is usually included in the form of studies or usability tests. As a rule, this takes place at the beginning of a project and is regularly checked over the course of the project. The co-prototyping accompanies the design process intensively and continuously. Relevant data is collected from users in the process and checked in a feedback loop. The participatory prototyping assumes a prominent place in the drafting process and extends across almost the entire process. The design itself is more for the shaping of the process and for an analysis format.

Design Prototyping Based on the Example of OLED Technology A design study[5] for OLED technology[6] was selected for the example of design prototyping.

[5]The study was prepared in a cooperative project with students in the design area of interface and interaction design at the Berlin University of Arts, the Fraunhofer Center for Responsible Research and Innovation and the Fraunhofer-Verbund Mikroelektronik (Fraunhofer Association of Microelectronics).

[6]The organic light-emitting diode (abbreviated as OLED) is a thin lighting element that consists of multiple layers of organic semi-conductor materials. In comparison to the conventional light-emitting diodes, these OLEDs can be produced less expensively since they can be applied to large areas of space in a special printing process. OLED can be used as an extremely thin panel radiator; the material can be transparent and flexible, produces high-contrast light with high colour

The results of this study were presented at the Plastic Electronics Conference 2011 as part of the SSL Semicon. For the study, a group of design students examined the possible potential of OLED technology and developed about two hundred ideas in an initial workshop on the possible uses of the technology in different markets. Design and technology experts evaluated these ideas in terms of their innovativeness and usefulness. The ideas that received the highest marks were refined in a two-week period for design prototypes. On the basis of the collection of two hundred approaches to using the ideas in various markets, it was possible to carry out an early qualitative and quantitative evaluation of the potential. For some markets, substantially more ideas were developed and were also given higher marks. This concentration of ideas can be considered the first indicator for examining a given market's suitability as a starting place. The two hundred ideas could be analysed in terms of used qualities, e.g. in regard to other material characteristics, shapability and supplementary sensor interfaces. Due to their qualitative information, the ideas transferred to design prototypes could also be evaluated with respect to the useful characteristics of OLED technology for central areas of application. For the field of application called "New forms of visualisation", which was developed in the workshop, the transparency and the flexibility of the material, as well as modular coupling mechanisms with integrated sensors are important, for example. The field of application referred to as "textile interfaces" assumes the generation of electricity on the human body and processing directly in the textile material, while the field of application referred to as "city periphery" is necessary for organic growth, as well as for durable and sustainable material. An example of a particularly innovative application was the self-growing organic OLED display. While the focal point to date in the development of OLEDs from a technological point of view has been the light and its properties, such as brightness, light colour and size of the display, the design perspective made it equally clear that, from an application perspective, it is important that intelligent interaction and communication occur with light. Not large displays but rather light as a means of communication, integrated in the central areas of life, is used. One might be inclined to modify the Ford quote mentioned earlier, namely, "Not larger, but rather modular, intelligent light". This example shows that the main usage characteristics can be identified in design prototyping in a very short period of time and at an early stage of technological development. However, there must be a relevant, qualitatively examinable number of design prototypes, and these can be prepared in an initial study, directly by designers. A specific evaluation tool for the monitoring of design prototypes for certain sectors and fields of technology would be an alternative approach. User groups are not directly included in this process. The results from the design prototyping can be discussed and evaluated with various user groups, which

(Footnote 6 continued)

quality, and the colour can change in the process. The material has a potentially long life and largely consists of environmentally-friendly materials. OLED is mainly used in screens and displays and for large-scale lighting at the present time.

may reflect possible requirements for future product development. The number of design prototypes should fall in a range between five and twenty-five. If fewer than five prototypes emerge in the process, this means that the technology has been developed too specifically, so a range in usage can no longer be developed. Conversely, a number above the upper limit of twenty-five design prototypes indicates that there are no significant differences in our various needs. While single, designer-developed approaches for possible product ideas initiated by individual designers are often gladly dismissed by engineers as non-implementable studies at such an early point in technological development, i.e. prior to the market launch, a larger and broader perspective in a qualitative and quantitative evaluation matrix for interaction and usage offers the possibility of integrating the requirements in the development of the technology at an early time.

Co-prototyping: The Example of Developing an Upper-body Orthosis An example of the co-prototyping format is demonstrated in the "Care-Jack" project, which involved the development of an upper-body orthosis[7] in the form of a jacket to reduce the physical burden on nursing staff in the strenuous nursing processes. This study was undertaken in the "Discover Markets" (cf. Schraudner et al. 2014b) research project.[8] For the study, typical situations in nursing care were re-enacted in a workshop with participants from active nursing care and developers of technology. Based on these detailed scenarios, it was possible to develop solutions for individual movement sequences, required action and communication situations in a co-prototyping process. To this end, scenarios in individual sections were replayed, whereby the participants took turns adopting the roles of patient and nurse. The process of co prototyping was carried out with a full-body protective suit from a home improvement store, which could be enlarged, labelled and changed with various materials. An accompanying questionnaire also allowed for specific feedback on the requirements for material, interaction, preparation and follow-up action with the orthosis. The participants used the "thinking aloud" method so that all aspects of the patient and the person acting as nurse were recorded. These original recordings, outlines and hand-written instructions as well as the prototypes themselves were evaluated for the study. Extensive information about the desired qualities of use in terms of material properties and interaction applications and communication applications became evident as a result of this evaluation.

One exemplary result of this study, first formulated on the basis of the co-prototyping—the need for a communication interface for upper-body orthosis

[7]An orthosis is a medical appliance or apparatus used to stabilise and guide movable parts and the upper body. The easily attachable, intelligent, active orthosis used here allows nurses reduce their physical burden by providing active support in lifting, carrying and transferring patients.

[8]"Discover Markets" is funded by the German Federal Ministry of Education and Research (BMBF), period 2010–2013 (funding code: 03IO1003). With "Discover Markets", a novel procedural model was designed to support the early identification of potential user groups' wishes and needs and the development of suitable business models for new technologies and product innovations ahead of the research projects. In the "Discover Markets" project, additional comparable co-prototyping studies were conducted.

that was not planned in the previous implementation—stands out in particular. Nurses identified the need to make it possible for the patient to indicate that a certain act by the nurse caused pain by using certain formats of tactile communication on the orthosis, for example. Tactile communication is faster than verbal and also possible for mentally impaired patients. Certain fine-sensor communication points on the orthosis also increase the range of possible uses. The example of co-prototyping demonstrates that, in technological development for specific areas of use and markets, it is possible to conduct studies with future users prior to the development of a product or service, if the co-prototyping is systematically tailored to the area of use (cf. Seewald et al. 2013). Accompanying documentation material that precisely describes the process in the greatest possible detail is important for systematic evaluation. In the process, it is preferable to include user groups that already have advanced knowledge of the process of the application area. The group should be diversified in terms of age, gender, experience and culture and can consist of experts, users and developers. The results of the co-prototyping provide specific information about the requirements in use and in the handling of new technologies in their area of use.

Participatory Prototyping: The Example of "Shaping Future", a Need-based Participatory Foresight Methodology The research project called "Shaping Future" (cf. Schraudner et al. 2014a)[9] was selected as an example for participatory prototyping since it developed methodological access for a participatory process to produce a technology preview that includes prototyping with non-experts (cf. Heidingsfelder et al. 2015). In a series of workshops, technology-interested non-experts are given the opportunity to anticipate possible futures in a method-supported way with the focal point of human-machine cooperation. At the centre is participatory prototyping, outlining the wishes and needs for interfaces in future human-machine cooperation in the form of prototypical implementations. These prototypical objects are not only interesting with respect to their specific recommendations for each case of implementation, but are also examined on a meta level for the technology preview: Which qualities do the used materials exhibit? Which sales paths are preferred? Will the prototype be used jointly and does it belong to anyone alone? Where does the energy come from? Should it be recycled?

In the participatory prototyping process, certain framework conditions are important so that no uncertainty arises with the participants. The environment, the rooms and the materials should encourage collective thinking, discussion and interaction. The materials for the prototyping can be simple: Cardboard and foam, everyday objects that are reused (everyday hacking), but also new prototyping tools such as laser cutters and 3D printers and technically-low-barrier interfaces such as the Arduino microcontroller.

---

[9]"Shaping Future" develops new methods for a participatory and need-based technology foresight. The results of the participatory workshops are analysed by experts and transferred to specific technology road maps. The project is funded by the German Federal Ministry of Education and Research (BMBF), preparatory phase: 2011–2012 (funding code 16I1630).

The process of participatory prototyping is broken down into individual steps that lead to the implementation of the object. In the form of a narrative integration (when/then, what for, why) of future human-machine cooperation, the first step is contextualisation. The participants then search for material that they want to use for their implementation. In this step, the material is not yet integrated into the object and is therefore more explicit in its statement with regard to the selected qualities. In the prototyping step, the sought material is transferred to an object context that describes the future interaction. The last step of the operating manual describes this narrative object from another perspective.

While the material search and the prototyping of the individual perspectives allow for leeway, the step with the operating manual causes the participants to change perspectives and to describe the use for other people: How does it start, how can it malfunction or be misused, how is the object disposed of? All the steps (the contextualisation, the material search, the prototyping and the reflection level) are evaluated with regard to the new research fields. The evaluation takes place on the basis of the qualitative data from the individual objects and descriptions and the metadata of the objects that can encompass between 20 and 100 objects in this process, and the participants' assessment in the workshop. The group of participants should comprise the greatest possible breadth of society so that the widest range of different backgrounds and perspectives can be included in the process of prototyping. The results of the participatory workshops are analysed by experts and transferred to specific technology road maps. These participatory road maps differ from the classical road maps, for example, due to their integration of previously not-integrated stakeholders and the outlining of new research questions and fields on the basis of requirements formulated by layperson (cf. Schraudner and Wehking 2012). This permits the inclusion of social perspectives in expert discourse. With this procedural model of participatory prototypes, non-experts can be actively integrated into research planning and the development of new technologies.

4 Summary

Potential of Design Prototyping for Future Research Planning and Technological Development Design prototypes reveal a lot about future uses and areas of use and application. If we look at the prototypes from various areas of design over a certain time period and consider them as a whole, we learn a lot about what economic, ecological, cultural, technical and social challenges were connected with them.

What added value does design prototyping provide for research planning and technological development? On the basis of the examples presented in this text, it is possible to demonstrate that design prototyping, co-prototyping and participatory prototyping can be important indicators for research planning and technological development. The respective format must be selected for the specific issue of technological development. In what particular form and under specific what circumstances daily life experts should be included or specific knowledge and shaping skills from design should be relied upon depends on their concreteness and applicability with regard to the technology to be developed.

Design prototyping provides the best results when a technology has been determined, but the user groups are defined as freely as possible. Co-prototyping, by contrast, should be built upon a set technological spectrum (different, but combinable technologies) and be open for various uses with a broad range of fields of application. In turn, participatory prototyping requires the greatest openness; neither technology nor application fields nor user groups should be limited so that an uninfluenced picture of the needs and wishes can arise in the future. These different requirements for each prototyping format also have an impact on the diversity of the results. Design prototypes are already firmly established in terms of their use and interaction proposals, and can be transferred to realisation. Design prototyping produces a picture of a later product by combining technical, use-specific and aesthetic requirements. The results in the co-prototyping provide descriptive and visual information for a technical realisation, but require in general another translation, interpretation and transformation to design prototypes. Only these can translate the various, partially contradictory wishes and ideas into various products and services that are logical, consistent and aesthetically appealing. Participatory prototyping requires not only an open process: This format produces least of all a picture of a later product; the prototype here is only to be understood as a process tool that provides information about future research agendas and technological developments in the concealed information. A translation, interpretation and transformation process in design prototypes is also possible in the process of participatory prototyping, but it requires a lot of feedback loops with the respective groups involved in the process in order to produce a picture of a later product or service.

Figure 2 demonstrates the various possibilities for use and application in each prototyping format.

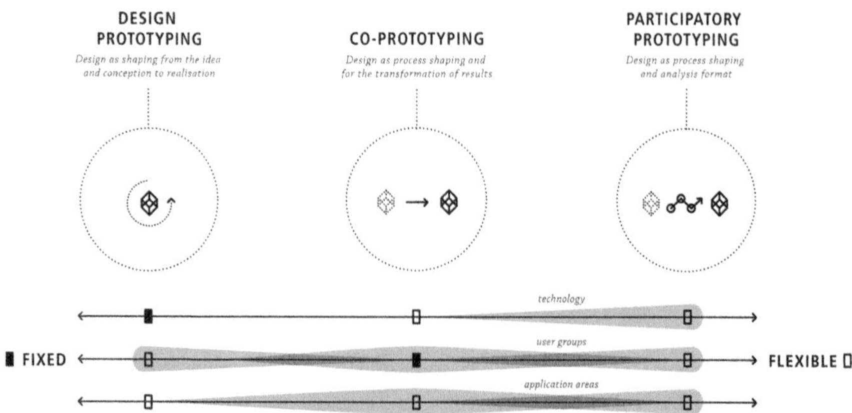

Fig. 2 User-centred prototyping for research planning and technology development

The research planning and technological development is part of the ongoing social and cultural development that can be designed with prototyping. The results from collaborating on research projects with designers, engineers and daily life experts demonstrate what the wishes and needs are for future technology.

As a result, this form of prototyping as an interdisciplinary form of communication can create a collective understanding of technological use. Open source products that offer interfaces for expanded use as open semi-finished objects could be conceivable. This might lead to product development, much in the vein of the open source movement in software development. Products and services would then have a basic configuration that can be expanded independently according to the user's personal wishes and ideas. Prototyping could then become a commonplace form of communication and interaction and become prevalent as a new medium in many areas of daily life, being used as a new form of education and training. This could certainly prove a fascinating challenge for the school of tomorrow if the old subject of "handicraft" enjoyed a renaissance through "prototyping".

Prototyping driven by design facilitates a discussion with respect to options for future actions. Thus, for future research, it is imperative that prototyping formats are defined to take a user's perspective into account as well. Prototyping driven by design can continuously integrate the user perspective into technological research. Consequently, design establishes itself as a discipline, within the technically natural, scientific research context.

References

Heidingsfelder, M., Kimpel, K., Best, K., & Schraudner, M. (2015). Shaping future—Adapting design know-how to reorient innovation towards public preferences. In *Technological forecasting and social change* (in press). http://dx.doi.org/10.1016/j.techfore.2015.03.009.

Schraudner, M., Seewald, B., Rehberg, M., Luge, M. K., & Kimpel, K. (2014a). *Shaping future*. Berlin: Fraunhofer.

Schraudner, M., Seewald, B., Trübswetter, A., Luge, M. K., Kimpel, K., Mizera-Ben Hamed, K., et al. (2014b). *Discover markets—Vorgehensmodell für eine Methodengestützte, nutzerorientierte Technologieentwicklung*. Berlin: Fraunhofer Center for Responsible Research and Innovation.

Schraudner, M., & Wehking, S. (2012). Fraunhofer's discover markets: fostering technology transfer by integrating the layperson's perspective. In D. B. Audretsch, E. E. Lehmann, A. N. Link, & A. Starnecker (Eds.), *Technology transfer in a global economy* (pp. 367–374). New York: Springer.

Seewald, B., Kimpel, K., & Schraudner, M. (2013). Discover markets—participatory methods for integrating user experience. In *Book of abstracts of the 1st European technology assessment conference "Technology assesment in policy areas of great transitions"* (pp. 171–172), (Prague, 2013).

Prototypes as Embodied Computation

Axel Kilian

Abstract The development of computational constructs that span the physical and digital realm opens up a new domain referred to here as embodied computation, a term introduced in my research at Princeton University. The role of prototyping is shifting from that of the confirmation of design assumptions in the early design stages to that of the embodiment of a design idea deployed into the world at large and continuously tested and updated digitally and if necessary physically throughout its lifetime. Feedback and control based on sensors and network-based information is enabling relatively simple mechanical structures to perform a wider range of tasks. There is a shift from mechanical complexity towards algorithmic complexity resultant from this change in many areas. In this article a number of prototypes are discussed in developing the concept of embodied computation through material and actuated constructs.

1 Introduction

The development of computational constructs that span the physical and digital realm opens up a new domain referred to here as embodied computation, a term introduced in my research at Princeton University (Johns et al. 2014). The role of prototyping is shifting from that of the confirmation of design assumptions in the early design stages to that of the embodiment of a design idea deployed into the world at large and continuously tested and updated digitally and if necessary physically throughout its lifetime. Feedback and control systems based on sensors and network information are enabling relatively simple mechanical structures to adapt to a wide range of situations (Mueller and D'Andrea 2012). Underlying this change is a shift from mechanical complexity towards algorithmic complexity that is observable in many areas from the long established fly by wire concept in aircraft

---

A. Kilian (✉)
Princeton University School of Architecture, Princeton, USA
e-mail: akilian@princeton.edu

C. Gengnagel et al. (eds.), *Rethink! Prototyping*,
DOI 10.1007/978-3-319-24439-6_4

design to more recent consumer products, such as smartphones. Yet the implication of these possibilities for more varied programs and forms such as architecture and engineering structures are less clear (Kilian 2006a). This article focuses on a range of computational factors in the exploration of design in small to medium design artifacts—in design, architecture, and engineering—and how the notion of the prototype is evolving and assuming different forms depending on the design challenge (Kilian 2006a). In this article, a number of prototypes are discussed with respect to developing the concept of embodied computation through material and actuated constructs.

Prototyping can be found throughout the design process; here I would like to focus on its role in different types of design exploration. Three different types of design exploration will be discussed: (1) the fine tuning of a design construct that incorporates known design parameters, (2) the creation of a new design definition in experimental prototype iterations and (3) the discovery of novel design constellations in a constraint solver based software (Kilian 2006a).

In a design process, every new design state is prototypical; it tests the further specification of a design idea in a new rendition, in a new medium and in greater detail. Also with increasing fidelity of design tools such as CAD software and rapid prototyping, the meaning of prototyping increasingly shifts to the further specification of the design and less so its traditional physically based role of design testing.

The design development as the formation of an idea in stages, through the translations from one implementation to the next, translating an idea into a different medium such as a sketch, then a text, then into geometry, then a physical construct and finally into a functional evaluation. Each translation opens up apparent gaps in the process and requires some of these gaps to be filled in order to accomplish the translation. Prototypes play a crucial role in this process, both as physical and algorithmic constructs and any combination thereof. Next, in greater detail, we discuss a number of these design explorations through prototypes.

2 Case Studies

Assembly Based Form, Prototyping as a Fine Tuning Exercise—The Plywood Chair The chair experiment (Kilian 2006a, b) was developed to demonstrate the possibility to create a plywood assembly from flat sheet, laser cut parts that achieve their curved state through cold bent spring loading during assembly, without glue or fasteners. To achieve this completion of the form through material based computation, the material response was tested through a series of fine tuning prototypes and the translation of the design intent into parametric proportionally flexible geometric models that represent and implement the desired curvature in a relational computational constructs. Those geometric constructs specified the laser cutting geometry, which then through spring loaded assembly of the parts induces the cold deformed curvature of the parts in the final chair. The creation of the chair is only possible

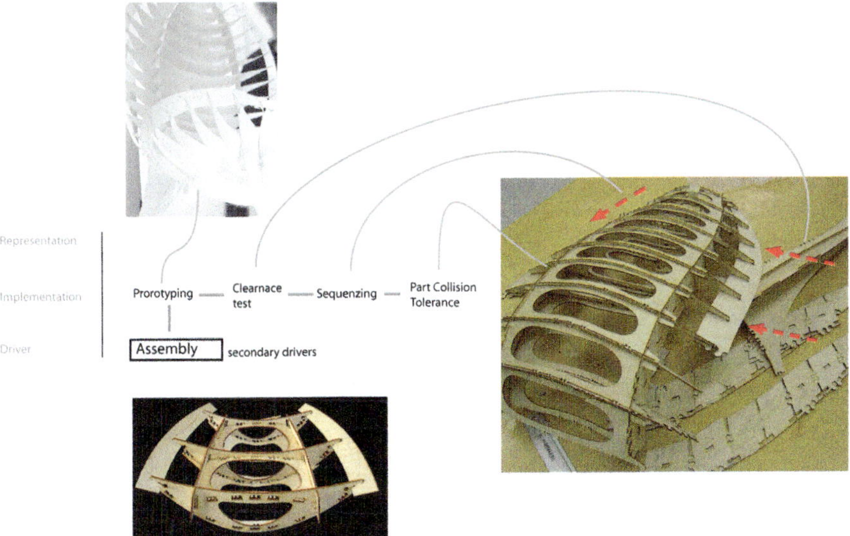

Fig. 1 Prototype series bent plywood assembly. © "Collection FRAC Centre, Orléans", (Kilian 2006a)

through the combination of geometric instruction sets generating the fabrication paths and the physical interaction and deformation of the material pieces to induce the final shape during assembly. Although the deformation is represented geometrically and implemented through NURBS surfaces, the final form is determined by the material interaction and the interaction of the assembly parts (Figs. 1 and 2).

The material behavior and tolerance parameter were explored through a number of prototypes of increasing complexity up to the full implementation, all based on an evolving CATIA parametric model. The translation of the intention into an assembly-based model occurred through a number of implementation and representation steps that were fine-tuned through physical prototypes. The resulting parametric model allows for the proportion and relational variation of all parts and through visual feedback the confirmation of the state of the model.

The adjustment of the parametric model allows for the regeneration of all joints and assemblies to create another chair geometry with different proportions. Successful outcomes are not guaranteed for all settings in this stage of the development and the two full prototypes created still exhibited detail flaws and are not reliable structurally overall, which would limit actual use. Here, however, the focus of the exploration was on the material formal interaction for an assembly test in a chair dimension. More development iterations would be necessary to achieve an everyday, usable chair.

Sequential Prototyping for Design Definition—Concept Car Exploration In this design exploration, the prototype played the role of defining the design space itself iteratively through the expansion of the design criteria with each successive

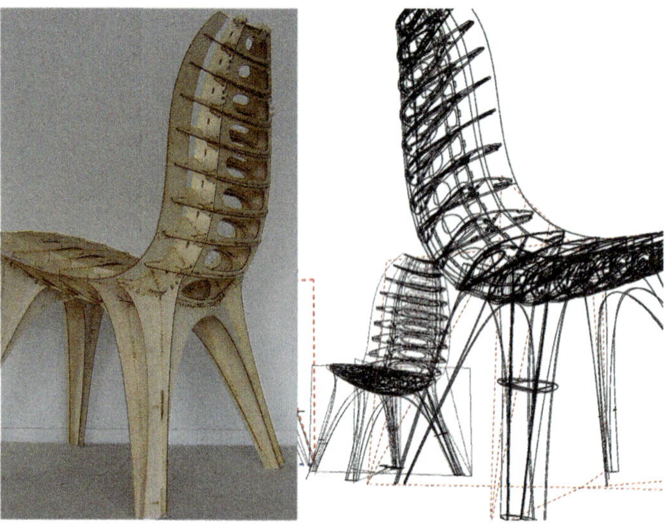

Fig. 2 Physical material based spring-loaded state versus geometric construct. © "Collection FRAC Centre, Orléans", (Kilian 2006a)

prototype. The evolution of the understanding of the design task is reflected in the prototype series. The embodiment of the design idea is also developed in the integration of actuation mechanisms that expand the design space, from the formal criteria to that of movement and flexible enclosures. Ultimately, the design cumulates in an exoskeleton-like extension of the body and was tested in a selective physical prototype that was an implementation of a selection of key components in a testable test rig with a meaningful interplay of features (Kilian 2006b).

The Prototype as an Idea-Defining Iteration Another use of prototyping is the establishment of a new design approach and test the effect of the partial implementation in a deployed scenario. The prototyping cycles then can lead to a successively more refined definition of the design task informed by the insights gained from the partial prototypes along the way. This process is well established but here the emphasis is explicitly on the exploration of a design idea through partial prototypes (Fig. 3).

The question of how to generate novel instances for an established design space could be approached through the development of a design language. In this case, the starting set was a set of writing devices that are analyzed for shared traits, such as the material used to leave a trace, how the handling is done and how the material is transported to leave a trace. A series of questions were developed as a starting point for the creation of a new instance from the common features identified in the analysis. This is an example of a very simplistic design and not representative of the open-endedness and complexity of larger architectural projects, but the study is a reminder that little progress has been made in the development of computational support in the concept forming stages of design. Most advances are situated in the

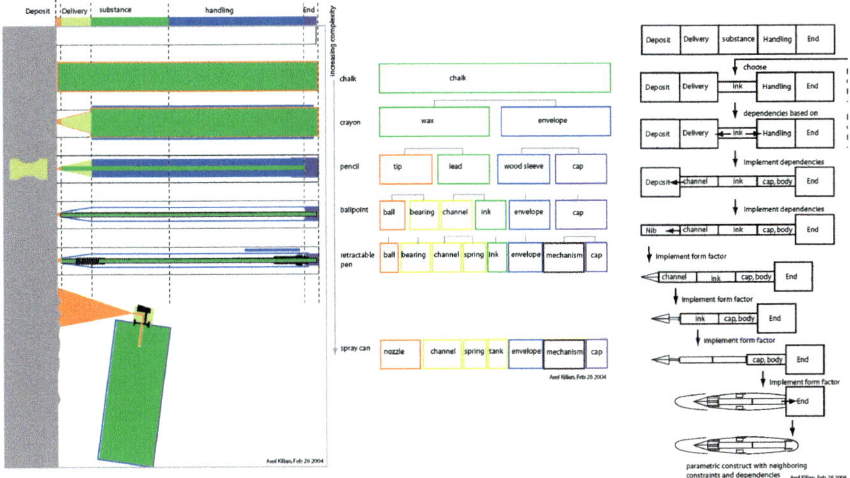

Fig. 3 Prototyping an idea process (Kilian 2006a)

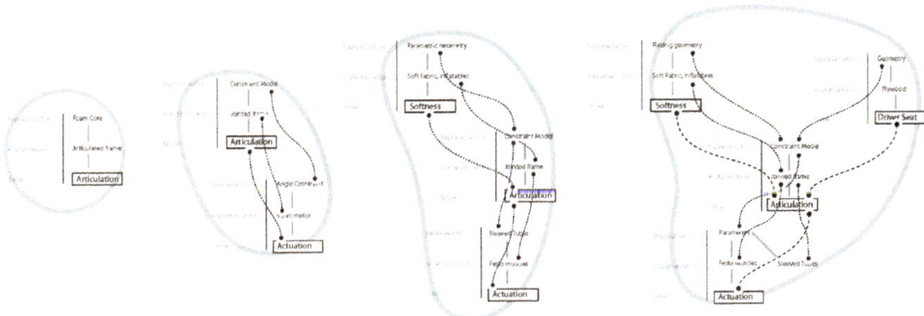

Fig. 4 Expanding design definition through prototypes (Kilian 2006a)

geometry management portion and the fabrication part of the design process. The following example of a concept car study conducted by the author with the William J. Mitchell smart cities group at the MIT Media lab (Kilian 2006a).

The design scope expands with each prototyping iteration and the definition of the design is becoming more detailed (Fig. 4). Here, prototyping is the successive definition of the design idea, not just its confirmation. The series of design studies are used to identify additional design features that are then added in the next iteration, beginning with an articulated chassis, as illustrated in Fig. 5. The degrees of chassis freedom demonstrate the obvious need for actuation, which comprises the first addition (Fig. 5).

The manual movement of the articulated chassis allowed the testing of the range of motions and the next iteration added several degrees of freedom, which no longer made it possible to control all six degrees of freedom simultaneously. The need for

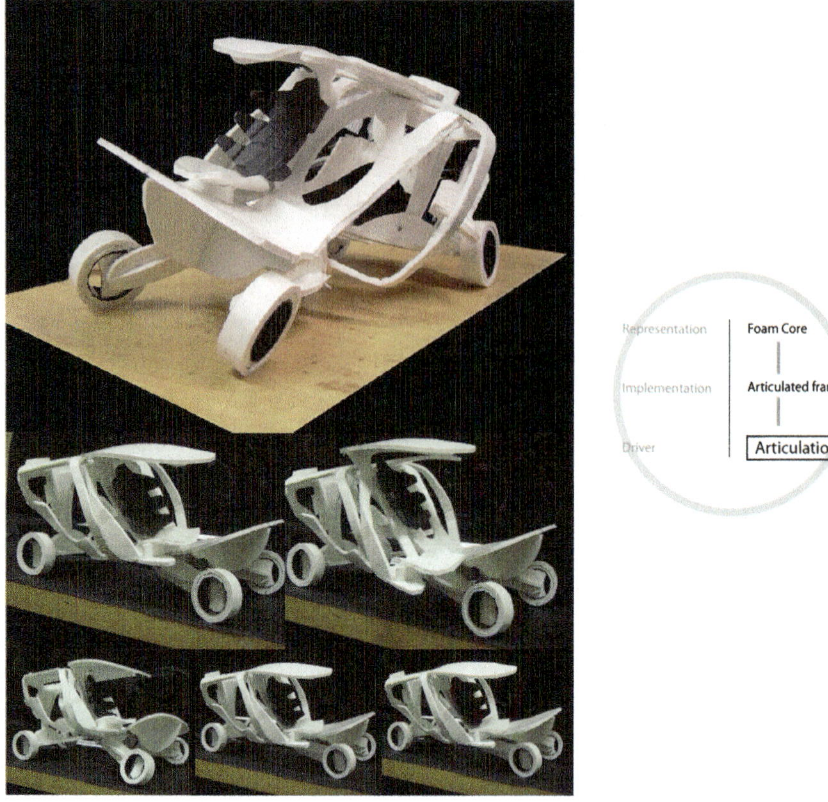

Fig. 5 An articulated frame as the starting point of a concept car exploration (Kilian 2006a)

simultaneous control triggered the introduction of servos and a micro controller to coordinate the range of motion and allow for design iterations through programming the controller and exercising the physical impact (Fig. 6).

Adding actuation to the chassis expands the design space with the movement choreography of an articulated body. Programming movement patterns allow for the exploration of the design space of motion as design expression. The insights gained from the six degrees of freedom test lead to the addition of two more degrees of freedom and a stiffer chassis in the form of an aluminum waterjet cut assembly (Kilian 2006a) (Fig. 7).

The next iteration included the development of the human machine interface in the form of an exoskeleton seat that snaps on the human body and maps the human motion onto the car chassis. The construct becomes a wearable extension of the body, covered in a soft adjustable skin and held together by a skeleton-like chassis with pneumatic actuators. A full-scale selective prototype was developed in order to make it possible to experience the concept physically and develop the interplay of parts. The idea of a selective prototype is to include all crucial elements in a

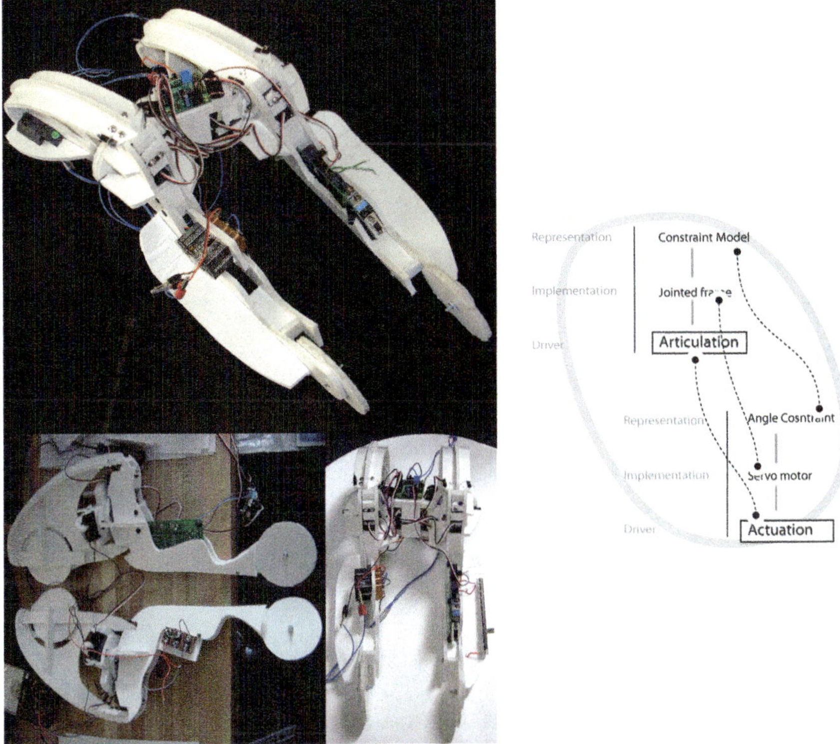

Fig. 6 A microcontroller servo actuated frame for movement prototyping (Kilian 2006a)

meaningful constellation, but at reduced complexity, in order to manage the cost and scope of construction while still allowing for the experience of the effect. Figure 8 shows the design iterations through progressively more detailed and designed physical iterations, from cardboard to milled foam to carbon fiber construct.

The full scale selective prototype shows the final design by the author with bent plywood seat as implemented by Patrik Künzler and Enrique Garcia, along with the carbon fiber integrated suspension wheel by Peter Schmitt and chassis by the author and Peter Schmitt (Fig. 9).

Prototyping Interdependencies—Form-Finding Application for Design Discovery When the prototype becomes a programmed construct, it can function as a dependency prototype that allows for the exploration and discovery of novel design solutions within a defined set of constraints. In this case, the example is that of a form-finding hanging chain modeler that enables the user to set up the connection topology and the rest length of the geometry as well as material resistance in form of the spring constant. Simulated gravity then acts on the particle mass to move the geometry incrementally towards an equilibrium state. With respect to the form-finding application, these are states of equilibrium (Kilian and Ochsendorf 2005). Part of

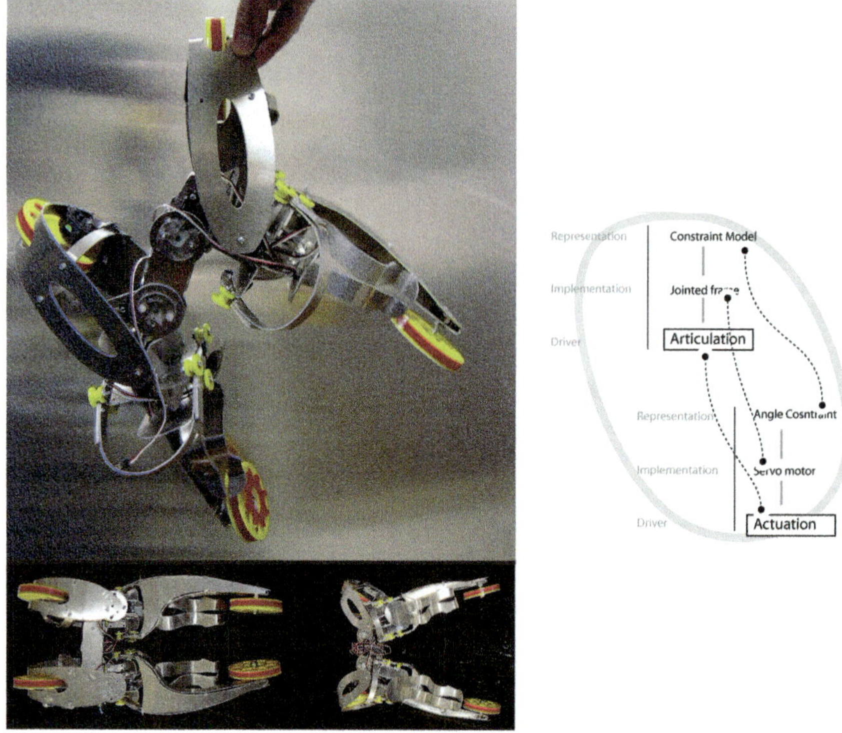

Fig. 7 Further development of the actuated frame with eight degrees of freedom (Kilian 2006a)

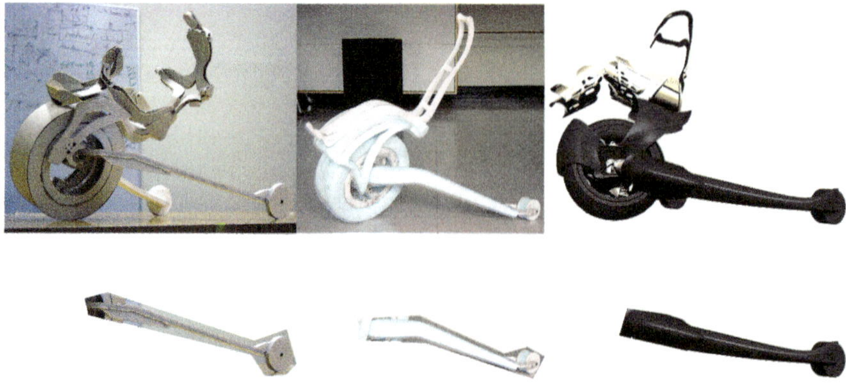

Fig. 8 Prototype iteration through increased fabrication precision and parallel design refinement (Kilian 2006a)

design shifts from a descriptive and generative operation of the intent into the playing of the relational construct more like an instrument in order to discover novel design constellations. In its extension, the form-finding becomes the steering of form, an

Fig. 9 Final selective prototype in carbon fiber, Axel Kilian, Peter Schmitt, Patrik Künzler, Enrique Garcia (Kilian 2006a)

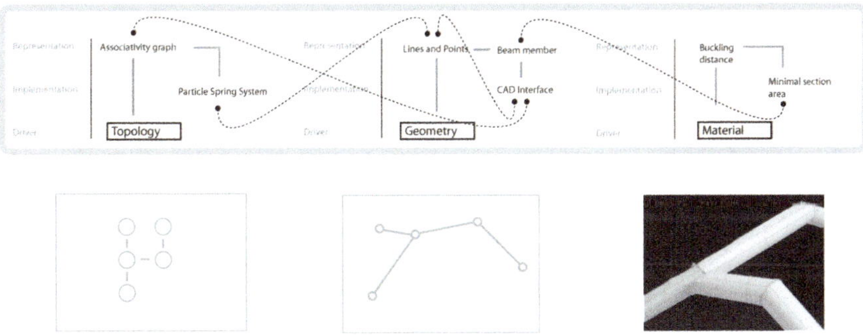

Fig. 10 Dependencies diagram of constraint prototype by programming (Kilian 2006a)

approach that embraces the open-ended nature of the definition of the dependencies as the design exploration unfolds, encouraging potentially competing factors to be leveraged against each other within the same set of constraints. For instance, in the force equilibrium example, this can mean extending the freely linked chain model that is purely in tension with moment-actuated joints to enable moment resistant sections as well. And more generally speaking, the design approach is the interplay between creating design constraints and exercising those constraints for exploring the design possibilities within the remaining degrees of freedom that still fulfill the desired dependencies (Fig. 10).

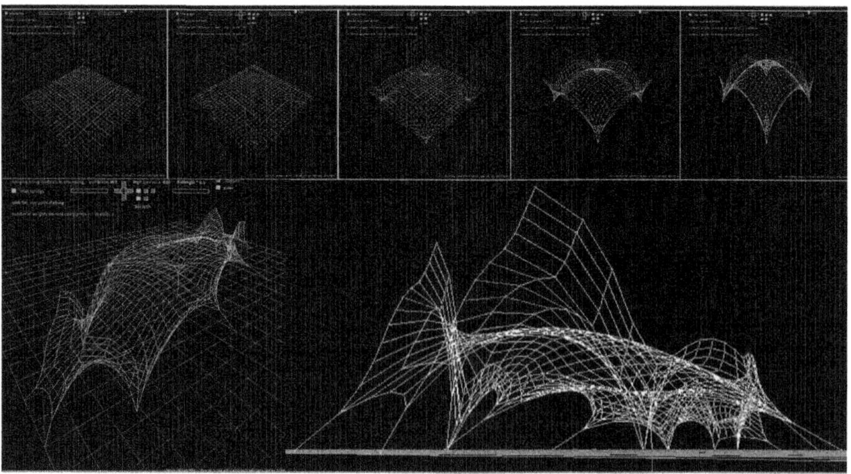

Fig. 11 Design process sketch in an equilibrium based form-finding tool (Kilian 2006a)

Design extends from descriptive geometry modeling to system modeling and the exercising of that constraint system. This challenges the notion of the design process because the interplay of design factors changes the setup and evaluation criteria of the process with different criteria, such as a state of equilibrium, compete with more established selection criteria, such as aesthetics (Fig. 11).

Embodied Computation Prototype—Active Bending Bow Tower Example A sensor-equipped and actuated structure serves as a small-scale test platform for exploring the range of posture changes that an active bending-based actuated tower can take. The goal is to develop a vocabulary of actuations to resist external forces and also use movement and shape change for design expression. Ultimately, this approach enables extending the prototype phase into the deployed design as it is possible to continuously update programmed behavior and, more importantly, to learn from the structure's environment (Fig. 12).

In smart phones and computers, the continued software update of devices is already part of the everyday cycle of product use and development. In the automotive sector, Tesla has repeatedly remotely updated the features and capabilities of their customer-owned cars by means of software updates. Due to their relative longevity, individual buildings and cities are promising candidates for retrofits of feedback and control systems to enable a more flexible response both to local and global changes. This is currently occurring already on a small-scale in networked climate-control systems, but it is certainly possible to imagine this approach extending into changing architectural programs and flexible, short-term use of existing structures. As the majority of the built environment will remain, retrofitting buildings and infrastructure (wherever possible) with feedback and control abilities

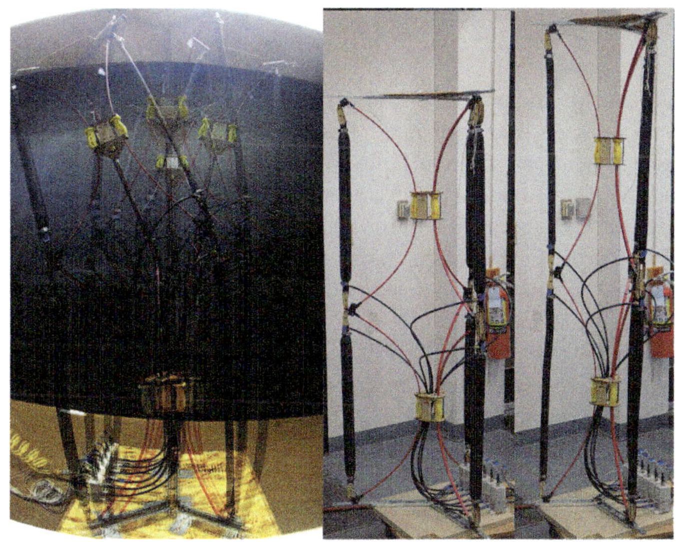

Fig. 12 Active bending combined with Arduino controlled actuation and sensor based feedback as embodied computation

may enable the existing structures to behave in ways that have positive effects on resource consumption and enable more flexible programmatic use, both collectively and individually.

Prototyping Material Organization—3D Printed Material States The printing of a material effects, i.e. printing with a material that is not different based on its chemical composition and that is not an assembly of parts, but rather exhibits different properties based on the specific, physical organization of the material require a different approach for realization. For example, in an instance where the shared cellular focal point creates a transparency effect that moves with the viewer as he or she circles the object and simultaneously conforms to the superimposed boundary effects of an inner compression dome and an outer moment frame to counteract the horizontal dome forces (and also provide a level seating surface) due to the material organization details, 3D printing is the only feasible production technique. The design construct relies on the printing process to become a physical object due to geometric properties. The material embodiment through 3D printing is the only form of materialization and testing of the object. The process has its own constraints, such as limits in the maximum overhang distances and angles of the pieces while being printed upside down. Additionally, production time is a big problem, requiring over 400 h of printing, as well as printing in nine parts due to the limited print volume of the simple desktop machine 3D printer.

3 Conclusion

Understanding of prototyping is evolving rapidly, particularly in its relationship to the design process. As discussed in this paper, due to the increased possibilities of linking design intentions into the built object, the separation between the design process and the finished artifact is disappearing on all scales. This presents unique opportunities for the continuation of the design process and the delivery of feedback from existing objects to the design teams developing them. Architecture presents a special case due to the relative longevity of its build constructs and the higher likelihood of reuse of existing structures. The prototypical experiments discussed here span a range of research interests under the general umbrella of embodied computation and represent ongoing research into the relationship of design, machines and the physical artifacts as created by the author.

References

Johns, R.L., Kilian, A., & Foley, N. (2014). Design approaches through augmented materiality and embodied computation, robotic fabrication in architecture. In *Art and design 2014* (pp. 319–332).

Kilian, A. (2006a). *Design exploration through bidirectional modeling of constraints*. PhD thesis MIT, Department of Architecture, Massachusetts Institute of Technology.

Kilian, A., (2006b). Design exploration with circular dependencies: A chair design experiment, CAADRIA 2006. In *Proceedings of the 11th international conference on computer aided architectural design research in Asia* (Kumamoto, Japan, 2006), (pp. 217–226).

Kilian, A. Block, P., Schmitt, P., & Snavely, J. (2006). Developing a language, for actuated structures. In *Adapatables conference*, (Eindhoven, 2006), (pp. 5–33).

Kilian, A. & Ochsendorf, J. (2005). Particle-spring systems for structural form finding. *Journal-international association for shell and spatial structures, 46*(2), 77–84.

Mueller, M. W., & D'Andrea, R. (2012). Critical subsystem failure mitigation in an indoor UAV testbed. In *2012 IEEE/RSJ International conference on intelligent robots and systems*, (Vilamoura, Portugal, 2012), (pp. 780–785).

Prototyping Practice: Merging Digital and Physical Enquiries

Mette Ramsgaard Thomsen and Martin Tamke

Abstract This paper examines the role of the prototyping in digital architecture. During the past decade, a new research field has emerged exploring the digital technology's impact on the way we think, design and build our environment. In this practice the prototype, the pavilion, installation or demonstrator, has become a shared research tool. This paper asks how this practice has formed by tracing the different roles of the prototype from ideation and design, to analysis and evaluation. Taking point of departure in CITA's own prototyping practice, we explore the relationships between physical and digital prototyping as a particular means of validation and verification. Here, a breadth of physical prototypes take on varying roles, in turn informing, testing and proving the research enquiry. The paper addresses how we can differentiate between these modes of prototyping and how.

1 Introduction

In 2002, at the Venice Biennale, Greg Lynn exhibited the culminating work on his project, "Embryological House". As part of this project, he exhibited a large model. Vivid blue and solid, it was larger than the body, suggesting something that seemed like full scale. I remember seeing this project and wondering why it was so large. It was neither a model, nor an installation, but rather something in between. It didn't present a spatial interior or a material logic. What was it about this amorphic project that necessitated such a large-scale representation? And how did the sheer scale of the model itself enable a new experience of effect?

M. Ramsgaard Thomsen (✉) · M. Tamke
CITA, Centre for Information Technology and Architecture, KADK, Royal Danish Academy of Fine Arts School of Architecture, Copenhagen, Denmark
e-mail: mette.thomsen@kadk.dk

M. Tamke
e-mail: martin.tamke@kadk.dk

C. Gengnagel et al. (eds.), *Rethink! Prototyping*,
DOI 10.1007/978-3-319-24439-6_5

Shortly after this, we started our own 1:1 practice in CITA, Centre for IT and Architecture at the Royal Danish Academy, School of Architecture. Through installations and demonstrators, we found ways to query the realisation of a digital architecture as it interfaces with advanced programmable design tools and digital fabrication. In difference to Greg Lynn's blue model, the inquiries move beyond the purely representational. The full-scale material investigation enables a searching of the spatial, structural and material logics of a research enquiry and is central in developing, testing and evaluating an idea. At the same time, however, they retain the model's ability to abstract architectural design space and allow an isolated inquiry without engaging in the full complexity of programme, site and environment.

Since then the practice of building full-scale prototypes has become a common tool in the emerging research field of digital architecture. Research pavilions, installations and demonstrators have become shared instruments in the exploration of how new structural and material systems can be realised. It has become clear that the realisation of these prototypes allow the further exploration of digital design logics. From design intent, fabrication, assembly and performance—the prototype is a means of testing and informing the digital vision. It is through this emergent research practice that the field and its technologies has been probed and creatively expanded.

This chapter asks how this practice of full-scale prototyping has informed digital design practice. With point of departure in the forming of CITAs own research practice, the chapter queries the development of the full scale experiment, its forms and roles in the exploration of a new design logic. A central focus lies with the dual emphasis on digital and physical prototypes. In digital design practice, the physical experiment exists as a result of extensive testing and prototyping in digital design models. In the following we pose the question: what is the relationship between the digital and the physical prototype, how does the digital inform the physical and how does the physical inform and interact with the digital?

2 Defining Prototyping Practice

Prototyping practice is ubiquitous in architecture. The idea of the artefact as first and foremost demonstrating an idea is core to the tradition of architectural thinking. Whether embodying the superiority of structural competency as in the Eiffel Tower (1889), the heroism of an arising ideology as in the Tatlin Tower (1919) or the style of a new century as in the Barcelona Pavilion (1929)—this tradition of large scale testing presents the artefact not as a complete architectural edifice engaging the breadth of architectural concern, but rather as bracketed by its particular investigation.

Prototypes are therefore singular in their ability to isolate an enquiry while at the same time communicating a vision for a new programme of architectural intent. They engage a hybrid territory in which they act as test objects as well as spatial probes. Different from models, they present a particular structural and material logic that is not only represented but also fully realised and embodied. Here the material

Fig. 1 The Dermoid demonstrator, 1:1 research by design exhibition, Meldahls Smedie, Copenhagen 2010

execution, as well as the methods and technologies of realisation, are achieved (Fig. 1).

At the same time, however, they present spatial entities that can be experienced directly and occupied in the same way as architecture. It is through direct spatial engagement that the prototypes become perhaps not architecture but *architectural,* probing and suggesting something larger than its technical, structural or material enquiry.

3 Prototyping in Digital Design Practice

In digital design practice the prototype occupies this hybrid territory becoming both the main means of evaluation as well as the central communicant for a new digital practice. It is through the large-scale demonstrators, installations and pavilions that the scope of these new design methods is conveyed externally to the broader architectural audience. But the prototype is also an internal tool of validating and the digital design model. As digital design practice expands its tools to include advanced simulation and links to an extended practice of fabrication and material creation, prototypes become important ways of ensuring the reliability of digital design processes. The physical prototype is here developed in parallel to a set of digital prototypes or models. But rather than understanding the process of creation as single paths leading from digital to physical, the physical prototype is understood as an integrated tool which tests and informs the digital.

In CITAs research practice, the investigation into digital fabrication has led to a focus on material performance. By understanding materials as neither static nor inanimate, but as engaged by complex behaviours and performances, our research questions how computational design methods can lead to new material practices. By designing for and with material performance, such as the bending of wood, the deformation of steel or the stretching of textiles, our aim is to expand structural and material thinking creating new lightweight, flexible and resilient structures.

A central interest, therefore, has been to understand, formalise and design dynamic material properties and employ these in the design process. This research investigation relies on the creation of new computational design methods that integrate material and structural simulation. The ability to model force and flow, to compute complex inter-scalar dependencies in advanced simulations and to interface these with intuitive design environments are fundamental tools in our practice (Nicholas et al. 2012). These models—or digital prototypes—allow the shaping of anticipations of how active structures employing the bending, stretching or deformation of materials behave. At the same time, the interest in material performance has resulted in a reliance on full-scale physical prototyping. Material performance is complex to represent and the traditions of scaled models are difficult to uphold when investigating material performance. In a scaled model, modelling materials inevitably differ in their behaviour from those that they represent. Working with material performance therefore means working with the real materials and therefore working in 1:1.

In our research on active bending structures, including Thicket, Dermoid, The Rise and Tower (Ramsgaard Thomsen 2011; Tamke et al. 2012, 2013), we differentiate between light-weight spring-based simulation tools allowing for the sketching of material intent and more solid FE analysis tools as means for testing and verifying design strategies. Both rely on the creation of material prototypes from which empirical data is gathered and used to inform design strategy.

As such, the practice of digital prototyping is different from other parallel efforts in neighbouring fields. Digital prototyping, or virtual prototyping, is a shared method of digitally designing and investigating a product or process (Petric and Malcolm 2003). The aim is to design, iterate, optimise, verify, and visualise a product as it is being created, often by a multi-disciplinary and large design team. Originating in the field engineering and industrial design, digital prototyping is seen as an alternative to physical prototyping and has fully replaced physical prototyping in some fields. Aviation industry companies like Lockheed Martin ceased using physical prototypes in the 1980s (Wong 2006); instead, the digital model has become the single place for conducting required verification, simulation and testing.

In difference to this exclusive use of the digital in other domains, it is exactly the interaction between digital and physical testing that is central to architectural digital design practices. In our projects, the use of prototypes, not only as a means of testing and verifying, but also for designing and creating designs, permeates the design processes. Physical prototypes are used to gather empirical data to inform and calibrate simulations. They are used to ideate connections and detailing while simultaneously understanding their structural performance. They are used to test mass customisation and map file to factory systems. They are used for checking assembly systems and sequences. Finally, they are used to test, evaluate and communicate overall structural and material performance and scope of a particular investigation (Fig. 2).

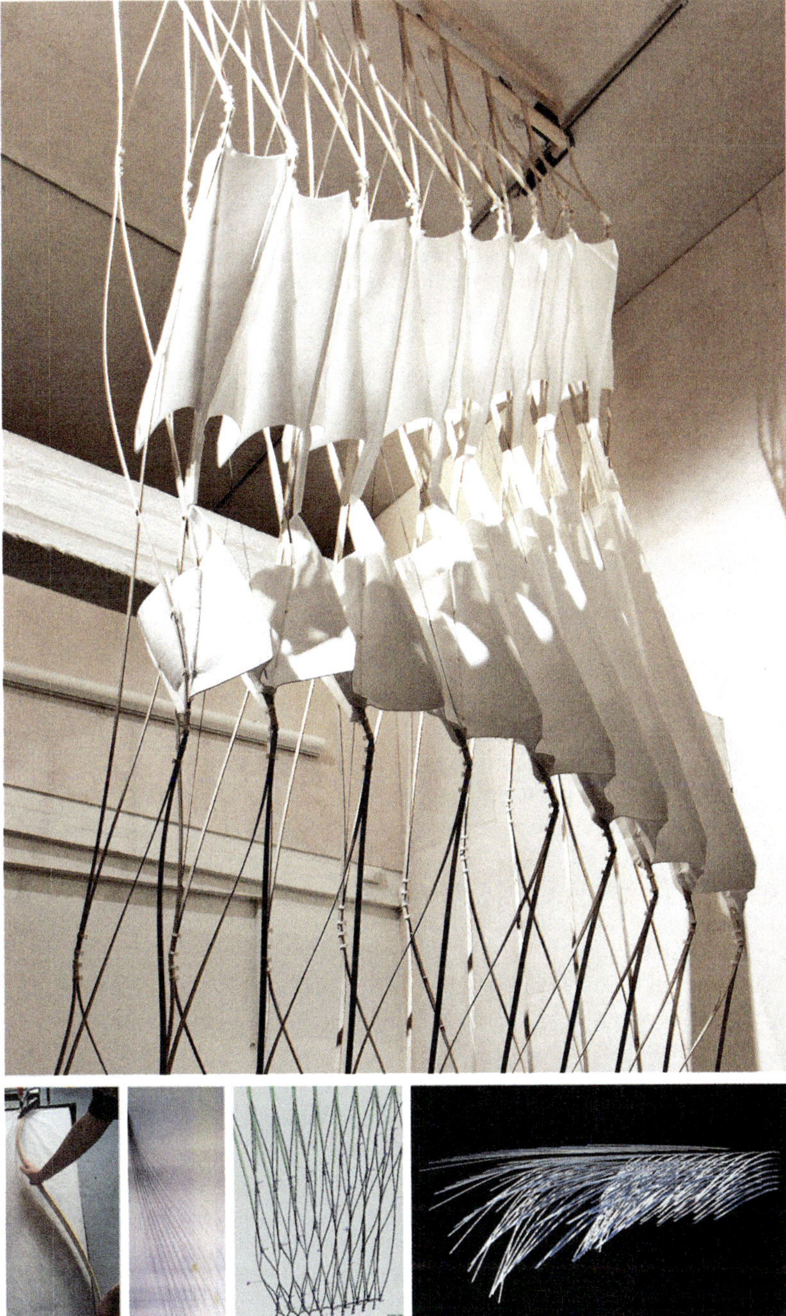

Fig. 2 Thicket and Thaw: material prototypes are used to inform digital design models of the formal properties of material deformation. Digital prototypes are used to simulate aggregate behaviour of structural assembly

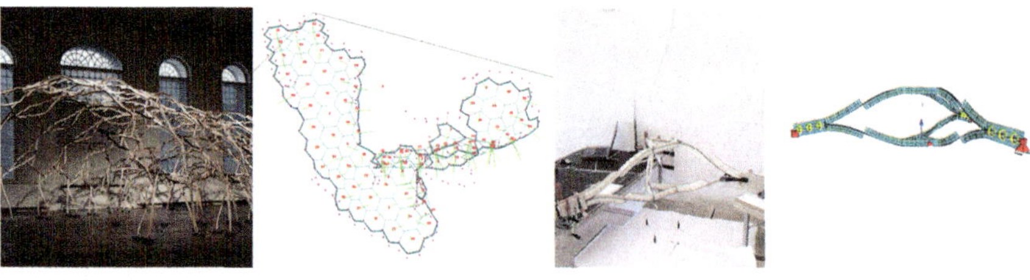

Fig. 3 Dermoid: the digital prototype as design tool and the physical prototype as ideator of material systems. Both inform high end FE simulations that are verified through comparison with a 3D scan of the final demonstrator

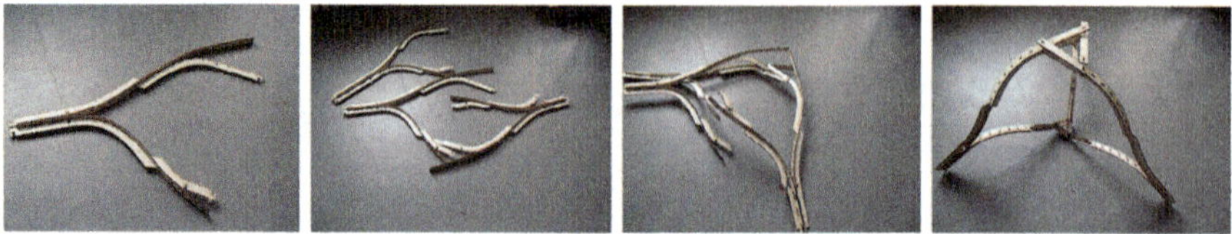

Fig. 4 Dermoid: physical prototypes ideation detailing of a double layered system of zipped flanges and webs. All connections are friction only

In Thicket and Thaw (Ramsgaard Thomsen 2011), digital and physical prototypes are used to inform the design of a pleated wall of ash slats. A primary prototype tests the twist and bending of individual slats deriving empirical data encoded into the parametric model. This allows us to design for and with the active bending of the material under force. The model is verified using light weight simulation tools (Maya Nucleus Engine) (Deleuran et al. 2011). Here, the aggregate behaviour of the slats in unison is simulated. Again, the digital prototype is based on empirical data gathered from physical prototypes (Figs. 3, 4 and 5).

In Dermoid, a collaboration with SIAL, RMIT and KET, UdK Berlin,[1] the interaction between digital and physical prototyping is expanded as an iterative process appearing across the design process. Light-weight simulation tools (Maya Nucleus Engine and Rhino/Grasshopper Kangaroo) are used to develop the overall design intent. Here, a digital process of inflation and relaxation of a design topology is used as the basis for understanding the distribution of elements. A further set of physical prototypes ideates and develops the detailing of the forking element. These are further tested in a set of parallel digital and physical prototypes that examine the performance of the fully scaled and realised element. The physical prototype is used to gather

[1]Dermoid is developed through the Velux guest professorship with Mark Burry at CITA in which the full teams of CITA and SIAL were involved. Further collaboration was established with Christoph Gengnagel's KET/UdK Berlin team to develop the FE analysis.

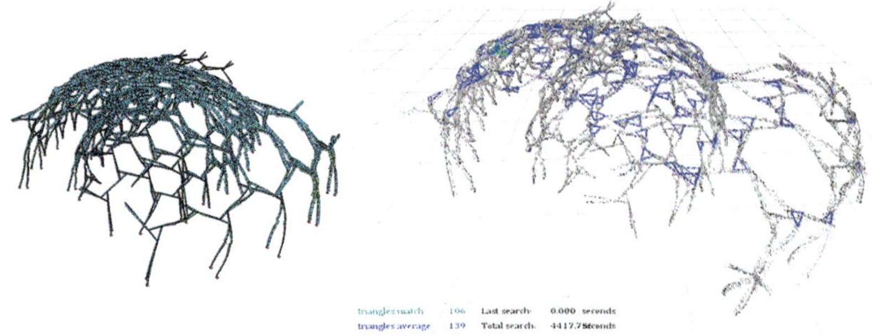

Fig. 5 The final evaluation of Dermoid took place through the writing of a feature detection algorithms that sequentially compared nodes in the 3D scan with nodes in the FE. Main deviations are highlighted in *blue*

data for the calibration of a detailed FE simulation (Sofistik), recreating the measured stresses within the element. Dermoid was built again for the Copenhagen Design Week and the Design Hub, Melbourne. Here, simulations of the individual elements are collated into an abstract overall FE simulation enabling an understanding of the overall performance of the structure. Finally, the resulting demonstrators were 3D scanned and used in a concluding evaluative digital prototype. An algorithm was developed to calibrate and compare the differentiation between simulation and realisation affording a solid understanding of the deviation between anticipation and outcome (Fig. 6).

In The Rise, initial physical prototypes were used to ideate the material system of connections, bundling and branching (Tamke et al. 2013). These were then used as input for a set of light-weight digital simulations formalising the geometric deformation of the material under load. In The Rise, geometry is achieved through the control of material behaviour. As elements branch we use "oppositional active-bending" to understand how the structure deforms. Here, elements in distinct orientations and of different stiffness work against one another to guide outgoing struts in their desired direction and the overall shape. A digital process for specifying member size and orientation is empirically calibrated through a tightly coupled series of physical investigations and a second set of prototype assemblies. These generate data about the rattan's bending performance throughout the structure and become crucial to the development of assembly logics, ultimately driving material specification within the digital model (Fig. 7).

In Social Weavers, a collaboration with Monash University and KET, UdK Berlin, initial prototypes are used to calibrate and understand the material performance of GFRP rods. Using different thicknesses, the aim was to formalise calculative models integrating both the performance of the single rods as well as their aggregate behavior (Stasiuk et al. 2014). The installation

Fig. 6 The Rise: the prototype as a means of formalising material behaviour informing the geometry of the simulated system

is conceptualised as a nest. It is comprised of multiple, actively bent splines that are articulated through a network of interwoven elements organised in distinct weave directions (Fig. 8).

The central component of the digital prototype is a custom-written, verlet-integrated particle simulation library that specifically allows for unfixed and transitional topologies. Throughout the simulation process, the designer can implement changes, additions and reductions to the topology and spatial configuration. The digital prototype further embeds assembly logic, self-specifying different rod thicknesses and their locations. A final demonstrator tests the resulting structure and its aggregate behaviour (Fig. 9).

In Stressed Skin, we explore the incremental forming of sheet metal. The project examines a very well-known material, applying well established practices for simulating elastic and plastic deformation while embedding it in a very unknown process of single point incremental sheet forming (SPIF). Physical prototypes were used to determine processing parameters and forming limits formalised as data for the computational model while digital

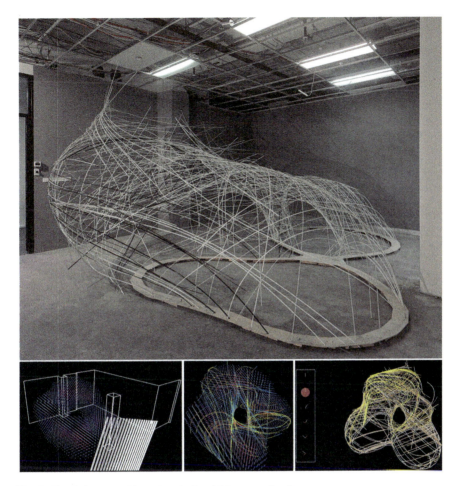

Fig. 7 Simulating assembly and assigning thicknesses of rods

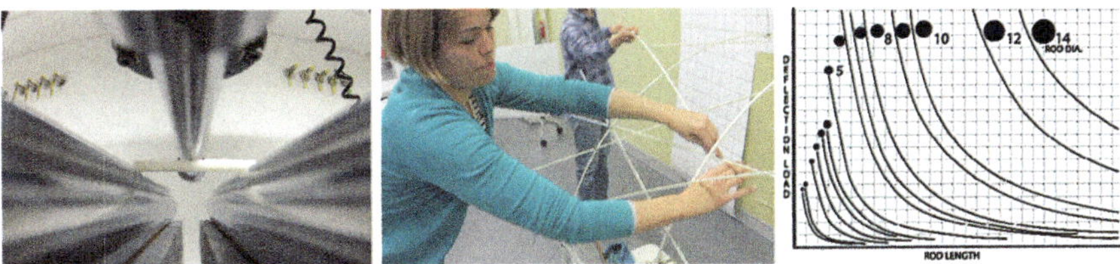

Fig. 8 Bending testing and prototyping connections with varying thicknesses of GFRP rods

prototypes enabled design within forming limits, the parameterisation of macro simulation with micro behaviour and the extraction of toolpaths for fabrication. A second set of prototypes ideates connections and understands

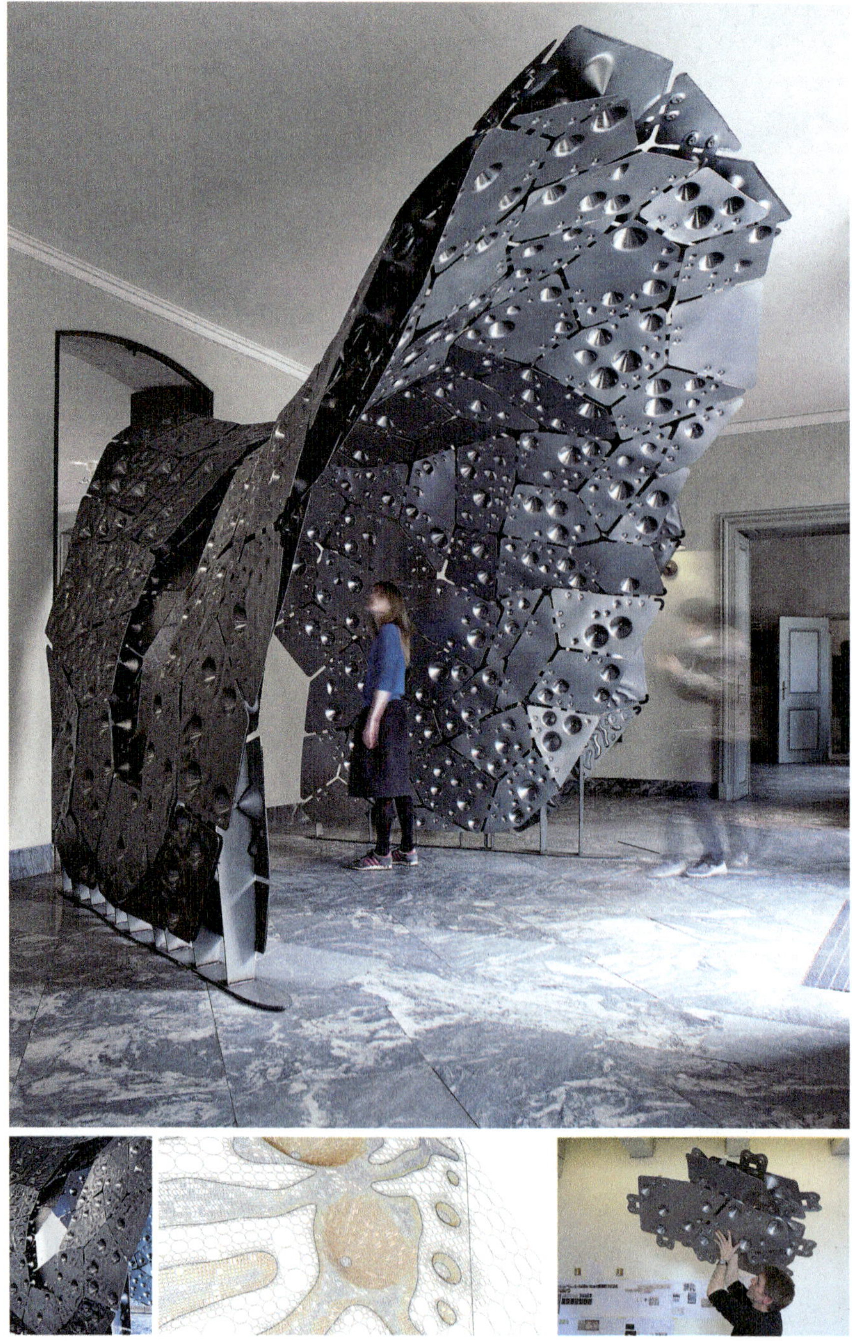

Fig. 9 Stressed Skin is a dual skin in which deep incrementally pressed indentations form the connections between the skins and further corrugation achieves stiffness

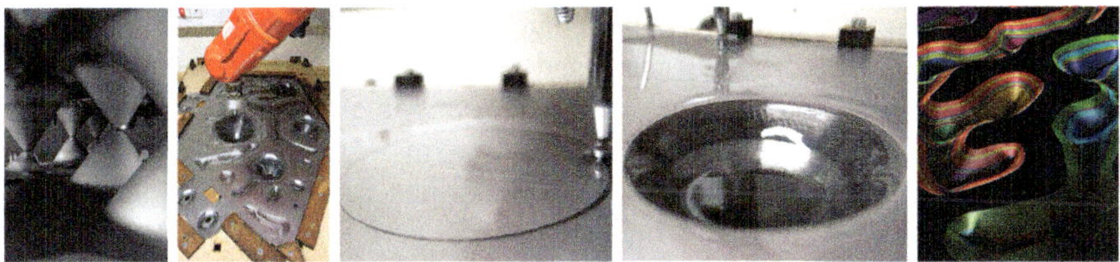

Fig. 10 Forming the SPIF indentations using an industrial robot

how the double surface of the structure performs. This is paralleled a simulation pipeline beginning with the generation of the panelling, a FE informed simulation creating the connections between the individual panels, a FE simulation of the performance of the overall structure and finally an FE based adjustment of the geometric stiffness of the individual corrugated panels. Other digital prototypes generate the assembly sequence of the structure (Fig. 10).

A final demonstrator is built to communicate the structural potential as well as the accuracy, calculation and control of the forming process and its associated digital design approach.

To understand this broad field of prototypes, we differentiate between different types of prototyping activity generating *material evidence* as testable design artefacts (Beim and Ramsgaard Thomsen 2011). The first distinction is between models and prototypes. Where models are understood as speculative and scaled, prototypes are fully scaled examining the realisation of an idea (Ramsgaard Thomsen 2009). The models is different to the prototype in that its core role is to invent design criteria, to question state of the art and speculate and theory build on the potential of new design methods and their technologies. They are an important part of the design process but fundamentally different to parallel activities of prototyping. As such, we follow the distinction made by Mark Burry in his paper "Models, Prototypes and Archetypes", in which he suggests a differentiation between the model that occupies a representational realm "generally in miniature, to show the construction or the appearance of something" and the prototype that is far closer to the realised. Here, the prototype is defined as "the original or model on which something is based or formed" (dictionary.com, quoted in Burry 2012). It is the alignment of the prototype with the original—or the *realised*—that is the central difference in our research practice.

The second distinction is between prototypes and demonstrators (Ramsgard Thomsen and Tamke 2009). In our practice, this distinction delineates the shift between preliminary test objects and final execution. In research practice, the idea of the realised is abstracted and the research enquiry results in the execution and realisation of a final prototype, namely, the demonstrator. Where the demonstrator concludes the research investigation and follows the consequences of design decisions, the practice of prototyping is much more divergent and fragmented. As an intermediate practice scoping ideas and testing concepts, techniques and technologies, prototyping searches the potentials of possible realisation. They exist as partial objects, tested in isolation on their own terms. As elements, details or test assemblies, they aim to discover the partial performances that can be assessed, and evaluated so as to use them to estimate the overall performance. The demonstrator conversely aims to set all of these divergent investigations into a common and concluding context. The demonstrator forces the research enquiry to engage with related processes of decision making as the architectural project. Rather than presenting an array of possible solutions, the demonstrator necessitates the prioritisation of one solution space over another in decision-making.

This emphasis on the design and implementation of material design experiments allows the research project to engage directly with the investigated techniques and technologies moving from design and analysis to specification and fabrication. This integrated approach—*research by design*—positions the research inquiries within a similar network of interconnected expertise and practice that make up architectural design practice.

4 Conclusion: Learning at 1:1

It is essential that we verify our models. When working with simulation, whether using light-weight, design integrated spring-based modelling, or more solid FE simulations, it is evident that the simulation is only as accurate as the data you enter. The legitimacy of a model informed by simulation can therefore only be the result of a meticulously evaluated process. Experimental research within the field of digital architecture probes the practices and materials of building. In simulating material performance, we rely on the merging of existing consolidated data and empirical data produced as part of the research process. How do we uphold the validity of these models and their data?

Perhaps an interesting way to understand how the evaluation of our model can take place could follow a scientific differentiation between verification and validation. In "Science in the Age of Computer Simulation", Eric Winsberg describes the scientific practice in which verification refers to *the mathematical exactness of a model*, whether it wields the results that a given equation proffers while validation refers to the *appropriateness of a model as a means of representing a given system* (Winsberg 2010). As such, verification and validation describe different levels of evaluation. Where Winsberg's point is to understand the interconnectivity between these two actions, and the blurring of an idealised line, the suggested terminology allows an important differentiation.

At CITA, the dual practice of digital and physical prototype is simultaneously used as a means of gathering data (for computation) and testing data (through computation). This testing—or verification—employs the data in a generative design context in which it constructs meaningful interactions with design intent and other input. In difference to science, the aim for architectural design practice is not only to analyse but also generate results. However, the "mode of testing" can be considered through the same duality offered by Winsberg. On one hand, we need to evaluate the precision and correctness of our models. This evaluation must necessarily include the accuracy of the data gathered through empirical processes. On the other hand, we must evaluate the *appropriateness* of our design models and their embedded generative logics. How do we question the idea of appropriateness? How do we define modes of probing, testing and evaluating the internal logics of our models? Is a model appropriate as soon as it provides the results we want?

References

Beim, A., & Ramsgaard Thomsen, M. (Eds.). (2011). *The role of material evidence in architectural research—Drawings, models, experiments*. Copenhagen: School of Architecture Publishers.

Burry, M. (2012). Models, prototypes and archetypes: Fresh dilemmas emerging from the file-to-factory Era. In B. Sheil (Ed.), *Manufacturing the bespoke* (pp. 55–74). Chichester: AD Reader, Wileys and Sons Ltd.

Deleuran, A., Tamke, M., & Ramsgaard Thomsen, M. (2011). Designing with deformation—sketching material and aggregate behaviour of actively deforming structures. In *Proceeding of Symposium on Simulation for Architecture and Urban Design, SimAUD 2011* (pp. 5–13). Boston, USA: Society for Modeling and Simulation International.

Nicholas, P., Tamke, M., Ramsgard Thomsen, M., Jungjohann, H., & Markov, I. (2012). Graded territories: Towards the design, specification and simulation of materially graded bending active structures. In *ACADIA 2012. Synthetic Digital Ecologies: Proceedings of the 32nd Annual Conference of the Association for Computer Aided Design in Architecture* (pp. 79–86). San Francisco: Riverside Architectural Press.

Petric, J., & Malcolm, L. (2003). Digital prototyping. In *Proceedings of the 8th International Conference on Computer-Aided Architectural Design Research in Asia, CAADRIA* (pp. 837–852). Bangkok, Thailand.

Ramsgaard Thomsen, M., & Tamke, M. (2009). Narratives of making: Thinking practice led research in architecture. In *Communicating by Design, International Conference on Research and Practice in Architecture and Design* (pp. 343–353), Brussels.

Ramsgaard Thomsen, M., Bech, K., & Tamke, M. (2011). Thaw—Imaging a soft tectonics. In S. B. Glynn (Ed.), *Making digital architecture*. Canada: Riverside Architectural Press.

Stasiuk, D., Nicholas, P., & Schork, T. (2014). The social weavers: Considering top-down and bottom-up design processes as a continuum. In *ACADIA 2014. Design Agency: Proceedings of the 34th Annual Conference of the Association for Computer Aided Design in Architecture* (pp. 105–121). San Francisco: Riverside Architectural Press.

Tamke, M., Lafuente Hernández, E., Holden Deleuran, A., Gengnagel, C., Burry, M., & Ramsgaard Thomsen, M. (2012). A new material practice—Integrating design and material behavior. In *Proceeding of Symposium on Simulation for Architecture and Urban Design, SimAUD 2012* (pp. 5–13). Orlando, USA: Society for Modeling and Simulation International.

Tamke, M., Leander Evers, H., & Stasiuk, D. (2013). Growing timber structures—venation algorithms as alternative approach to integrate design with constraints from material, tectonics and production. In *Proceedings of the International Association for Shell and Spatial Structures (IASS) Symposium 2013, "BEYOND THE LIMITS OF MAN"* (pp. 23–27), Wroclaw, Poland.

Tamke, M., Stasiuk, D., & Ramsgaard Thomsen, M. (2013). The rise—material behaviour in generative design. In *ACADIA 13. Adaptive Architecture. Proceedings of the 33rd Annual Conference of the Association for Computer Aided Design in Architecture* (pp. 379–388). Cambridge, Canada: Riverside Architectural Press.

Winsberg, E. (2010). *Science in the age of computer simulation*. Chicago: University of Chicago Press.

Wong, K. (2006). What grounded the airbus A380? In *Cadalyst Online*, (Dec. 2006). http://www.cadalyst.com/management/what-grounded-airbus-a380-595. Accessed July 08, 2015.

Prototyping the Unfamiliar: New Dilemmas of Scale Within an Evolving Digital Design Landscape

Mark Burry

Abstract Designing spaces that are entirely unfamiliar in terms of cognitive spatial arrangements present particular difficulties for architects in cases where floor surfaces are not level, for example, walls are not vertical, and ceilings richly sculptured. One such space is presented here as an example of new dilemmas of scale that architects face—the 'Sala Creuer' above the crossing of the Sagrada Família Basilica. In such situations the only prototypes that can fully reveal the designers' intentions are full-scale mock-ups, or more typically, the completed built space. Scaled prototypes have other important roles to play especially within a rapidly evolving digital design landscape, but offering the end-user a credible preview of the anticipated spatial experience entirely unfamiliar in cognitive spatial terms is probably only a remote possibility.

1 Introduction

In this Sect. 1 shall offer evidence directly from architectural practice to support the contention that certain design aspirations may only be definitively prototyped as the final outcome, and that such a prototype is a mock-up, not a prototype unless we are talking about the completed work—the ultimate prototype perhaps. A major difference between architecture and product design points to every building being a prototype (for the next building): that the full lessons from testing and discovery only come when we experience the spaces created at 1:1. Obviously this is the case for any building design given that the outcome has been specifically set-up as something that has never been experienced before. This is why any copy of an existing building simply placed somewhere new is just that: a copy, and not a new design. Nevertheless as the design progresses architects use every means at their

M. Burry (✉)
Melbourne School of Design, Faculty of Architecture, Building and Planning, University of Melbourne, Melbourne, Australia
e-mail: mburry@unimelb.edu.au

C. Gengnagel et al. (eds.), *Rethink! Prototyping*,
DOI 10.1007/978-3-319-24439-6_6

disposal to test their evolving designs through prototyping, and with the new digital tools evolving at a remarkable rate, the architect has been able to offer more detailed confidence inspiring insights to clients than has previously been possible. With the appropriate equipment architects can offer clients glimpses of the future project through immersive 3D, for example, or even augmented virtual reality (AVR).

The same advanced digital workbenches that afford more insightful testing along the way through novel approaches to prototyping also challenge the architect to push their designs to offer innovative architecture redolent with vigour, expression and formal boldness hitherto never experienced. The more powerfully the digital aids contribute to designers' conceptual adventures, the more radical the new architecture offered to society. But with novelty and complexity comes the risk that what is being revealed has never been remotely experienced before, and that regardless of the apparent veracity of the augmented virtual reality experience, only a loose facsimile of reality is actually being proffered. To an extent AVR relies on our cognitive abilities being linked to our memories, and for the observer to comprehend what is being observed in order to interpret and evaluate the novelty within the design proposal. Essentially, there is a paradox here: better prototyping drives the designer towards producing designs that evade effective spatial and cognitive comprehension prior to the constructed object revealing all.

2 From Visualising the Unfamiliar to Prototyping the Unfamiliar

Putting this paradox upfront is an unhelpful opening gambit, of course, when we are looking at prototyping afresh, and at the exciting possibilities that the new perspectives on prototyping bring within that context. I am going to argue here that the paradox I have opened with is potentially a false one, in a sense, and I shall draw on recent revelations from the continuing construction of Gaudí's Sagrada Família Basilica in Barcelona to demonstrate that it is the exceptional nature of the architectural vision that is the key challenge to thinking about prototyping the spatial experience, and not necessarily the sensu stricto prototyping as tests along the way to production.

The design and construction of the space above the nave-transept crossing—the Sala Creuer (literally the 'Crossing Hall') straddles the digital divide as, on the one hand, it is a volume that Gaudí conceived in his final design for the building (1914–1926) (Fig. 1). On the other hand, the design development and construction has been undertaken entirely using a parametric digital workbench.

I shall outline and comment on the degree of usefulness of prototyping the design of this space with reference to the completion of the narthex over the Passion Façade portico to that same building.

Both areas are nearing completion at the time of writing in late 2015, beyond the time that this chapter is being put together. Curiously this before-the-fact

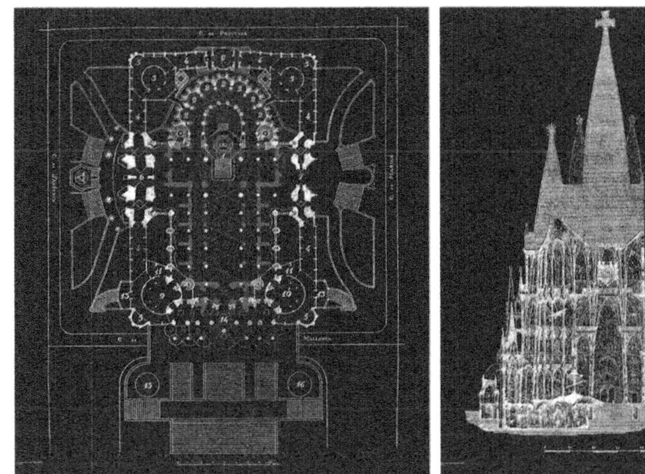

Figs. 1 and 2 Gaudí's original plan and section through the Sagrada Família Basilica

commentary suits this essay as my argument is very much about process and not about the whole design-to-tested-in-completion production cycle. The reason for this, briefly, is that it sometimes seems a more interesting challenge to record some unusual findings before-the-fact rather than reflect afterwards with hindsight. My principal contention here is that the unfamiliar cannot even be visualised meaningfully let alone prototyped. By meaningfully I mean that anyone following-up this almost completed work-in-progress account will have the advantage of being able to test my proposition that the absolutely unfamiliar cannot be properly understood until it is made familiar through experiencing the actual space post factum. This test is as true for the authors of the spatial narrative as it is for the reader. Photographs of the completed space cannot be furnished at the time of writing, just as this brief account of the challenges to test and reveal the qualities of the proposed outcome through a number of the following prototyping procedures cannot be affirmed as having been effective in providing an effective preview of the soon to be completed interior.

3 Design: Creative Pathway from Idea to Artefact

The following brief thoughts on design, even if neither self-evident in themselves nor necessarily universally held definitions, at least provide the glue to the following argument.

Fig. 3 Gaudí's original sketch for the Passion Façade (in 2015 the project is nearing completion)

As shown in the Gaudí's original longitudinal section through the nave, there is very little spatial information above the crossing as it has been entirely hatched-out (Figs. 2, 3, 4).

While there is not a lot of information for this upper section there are some crucial narrative overlays that propel the design towards spatial resolution consistent with Gaudí's more clearly stated design approach taken for the basilica interior. The first narrative—effectively the meta narrative, is Gaudí's use of the building as a whole to depict the life and times of Jesus Christ, and the wider Christian message of sacrifice, salvation, and glory. The central tower that sits on top of the Sala Creuer will be 172.5 m in height from the basilica floor level and will make this the tallest Christian place of worship ever constructed. In itself this detail may not be of particular importance, but the fact that Gaudí intended the cross at the top to be four-armed so that it will always appear as a cross when viewed from any direction is significant, because it is the climax of a vertical trajectory through the building for the intrepid. Being the tallest of all 20 towers this central tower is dedicated to Jesus Christ. Comparing Gaudí's highly specific rendering of detail for the interior of the basilica shown in the longitudinal section with that above the crossing, why has he has chosen not to elaborate the interiors of what are effectively service spaces above the basilica's main interiors?

Gaudí wrote nothing about his work during his 43 years as professional architect for the project so we have nothing to draw from there. Could it be that he regarded

Fig. 4 Close-up of the section with the hatched-out enclosed space above the crossing (Sala Creuer)

these service spaces as simply the humble voids above the elaborate interior vaults and beneath the necessary roofs and towers: rudimentary in nature? Or could it be that these spaces were so far down the track, and in themselves not contributing to the overall narrative of the building to any vital degree—such that their design and completion could be safely left to last, and without the need for Gaudí's own hand to any significant degree? From our perspective, Gaudí's posthumous successors, there remained the challenge of developing the sub narrative, that of the journey from terra firma through the firmament (the interior space of the Jesus Christ tower itself) upwards to the cross and the uninterrupted 360 degree views across the city that the four armed cross will provide the visitor. This journey up to almost 170 m has to be handled with care, as only a relatively small parcel of visitors will be able to entre the cross at the top at any one time. A 'collector' space is required, and the volume just above the crossing vaults is in the perfect position to provide such a facility. From a practical viewpoint this was an easy decision to make—Gaudí's section almost suggests it. From a liturgical position, however, this is rather an unusual situation, for the space is located 70 m above the altar below, and cannot form any direct part of the religious ceremony and ritual. It can offer some welcome rest and respite for the visitor almost half way through their arduous ascent to the top. For this reason the space has been designated as a small auditorium seating approximately 200 visitors, acting as a contemplative interpretation centre. It is assumed that a significant proportion of visitors to this space will feel that they have reached high enough.

The design process commenced in 2005 and coincided with the maturing of two significant innovations that the Sagrada Família Basilica design office had pioneered: shared 3D parametric models and rapid prototyping.

4 Prototyping Shared 3D-Models

The Sagrada Família Basilica architects studio began working with CAD software in the office in 1990, and coincides with the uptake of NC stonecutting. Ironically we were not able to find any architectural software capable of dealing with Gaudí's complex legacy resulting in the adoption of highly sophisticated software intended for the aeronautical industry. Equally ironically, because the NC take-up predated the adoption of CAD, far from being a paradigm of file-to-factory experimentation as it later became, initially the NC inputs had to be derived from manually drawn templates. The columns in the nave reflect this process. Adopting aeronautical 3D design software inadvertently led to the take-up of so-called 'parametric design' in 1992, at least two decades ahead of the more mainstream take-up by the architectural profession. Parametric design, or 'flexible modelling' as we preferred to call it, was the perfect digital aide-de-camp in the task of reverse engineering the surviving fragments of Gaudí's final design models made at 1:25 and 1:10 scales. As we moved into less charted areas such as the Sala Creuer space, being able to model flexibly through the use of such effective design modelling software proved to be invaluable.

The office employs rigorous CAD standards in terms of design documentation but in terms of the formal design each architect uses the software that suits them best. For the Sala Creuer design team we sketch modelled using Rhino™ and design modelled using Digital Project™. By sketch modelling I refer to sampling possible spatial configurations, which are brought to more tangible but parametrically variable fruition as design models. Aeronautical software was already set-up for teamwork and shared modelling a decade in ways that architectural 3D modelling software had but a limited capacity. I argue here that for this part of the Sala Creuer project the first prototyping innovation was prototyping the workflow of the shared 3D model (Fig. 5). We had to develop our own modus operandi in this regard, and lessons learned led to major spinoff research project 'Challenging the Inflexibility of the Flexible Model'.[1]

---

[1]A major research project funded by the Australian Research Council led by Mark Burry, Jane Burry, and John Frazer. Dr. Daniel Davis and Dr. Alex Peña de Leon were the postgraduate researchers at the project's core.

5 Prototyping Towards a Greater Spatial Understanding

Over the design period we were able to flexibly model in pursuit of two main design variants. The first is a potential schema in which the tower descended all the way to the crossing independent of the four lower 'Evangelist' towers that flank the perimeter of the Jesus Christ tower, and linked to them by a bridge across from each tower—visitors will ascend to the Sala Creuer via the Evangelist towers. Less obvious is an alternative schema whereby the exterior 'skin' of the Jesus Christ tower flares out to absorb the Evangelist towers in such a way that visitors can cross into the Sala Creuer under the cover of a vaulted structure above them. Variants of these two schemas are shown in Figs. 6, 7 and 8.

A most memorable decision point was the week when we were able to 3D print three such models and critique the possibilities—an unimaginable possibility five years previously when we first adopted rapid prototyping technology. It was not so much the fact that we could access 3D models so quickly; it was the fact that we could have them at all. The Sagrada Família Basilica has been blessed since Gaudí's day with plaster of Paris model makers supremely skilled in their craft. Not only would they not have been able to produce such finely detailed 1:200 scale models in reasonable time (taking into account that Gaudí's own practice had been to refine his design at scales of 1:25 and 1:10), rather than finding the entry of the 3D printer a confronting experience, the model makers studio were able to embrace their entry as an additional skill base within their repertoire. But how valuable were these spatial prototypes?

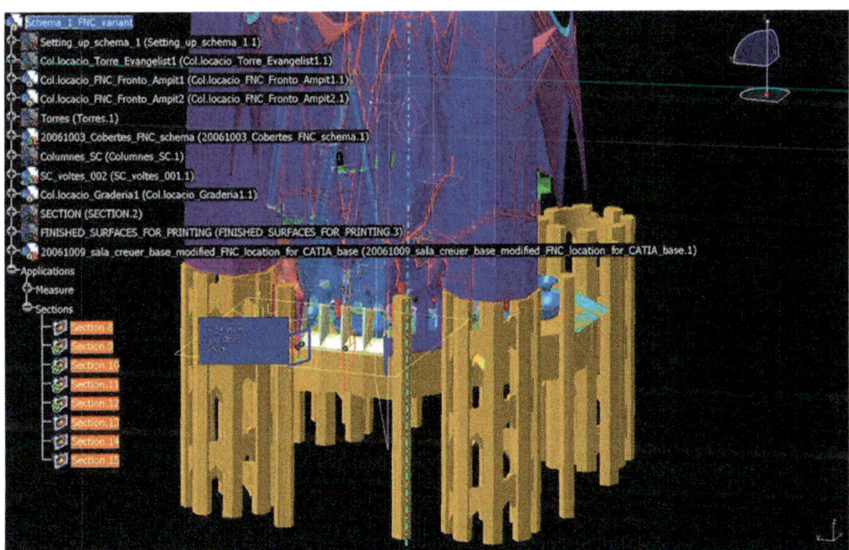

Fig. 5 Shared parametric 3D model—leading to the virtual 3D prototype

Fig. 6 Three early variants of the Sala Creuer schema

The answer is that they were very valuable in certain respects, but in other equally crucial respects—experiential proxies, for example, they offered very little. I shall digress via a brief account of an aspect of our shared 3D model workflow before returning to an assessment of the overall value of 3D physical and virtual prototyping for a project of this nature—a project characterised by there being no

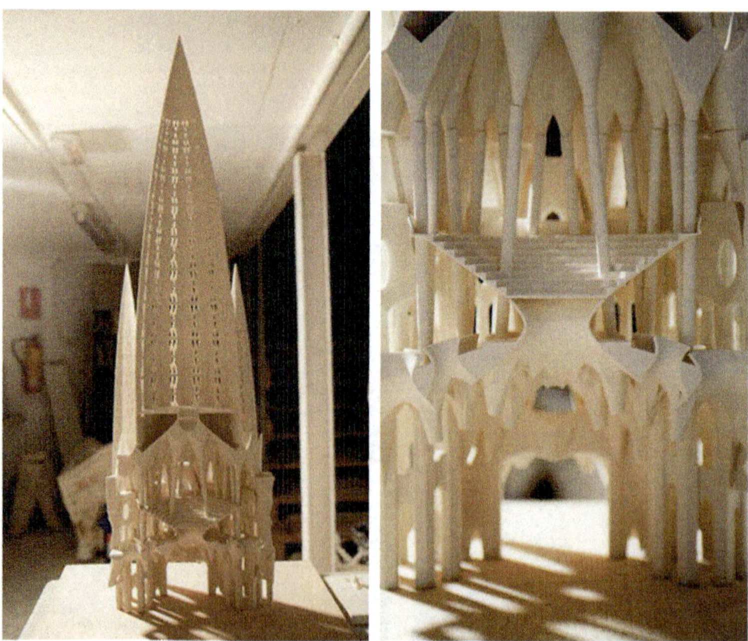

Figs. 7 and 8 General schematic of the Sala Creuer space as a cut-away 1:200 scale model and the detail of the chosen auditorium spatial configuration

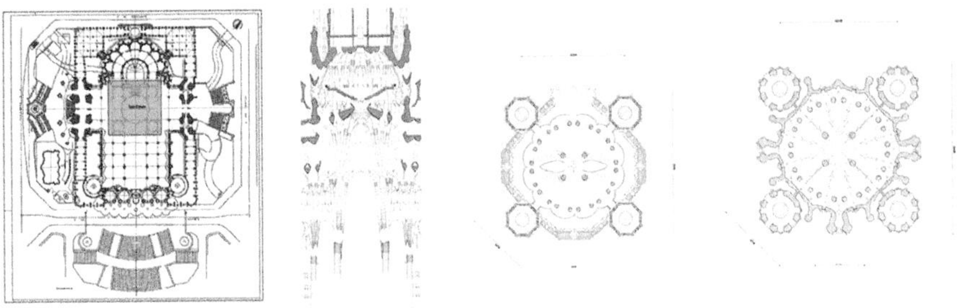

Figs. 9, 10, 11 and 12 Plans and sections of the Sala Creuer at various levels

known precedent for such a space as this, one with hardly any horizontal and vertical elements (Figs. 9, 10, 11 and 12).

The Sala Creuer is circular in plan almost to the point of being more Baroque than Gothic in nature.

With the exception of the seating and walkway accesses, almost all the surfaces are composed of assemblies of intersecting hyperbolic paraboloids and hyperboloids of revolution of one sheet articulated by planes. In combination the plans and sections offer insights into the subtleties of the spatial configurations but the experiential insights so gained are as limited as is that of interpreting Boromini's Santa Maria deghli Quattre Fontani in Rome from the drawings compared with visiting the interior. Curiously it is the model of the actual space within that particular building rather than a model of the building itself that offers a better interpretation through abstraction.

6 Prototyping the Unfamiliar

We tried 'visiting' a stereo 3D digital model in a CAVE which, beyond an initial 'Wow!' factor, left participants unsure about any of the typical spatially liminal attributes such as 'floor', 'walls', and 'ceiling'. Cognition of at least one familiar attribute would normally ground the spectator to frame the potential unfamiliarity of the others thereby offering some interpretative confidence—architectural elements of commonly experienced dimension and orientation, for instance, with which the brain can gauge more readily with the unfamiliar aspects of a spatial design through their juxtaposition with the familiar. With the Sala Creuer almost all of the spatial attributes are unfamiliar so even if one gets past the peculiarities of the configuration it is still very difficult to envisage the spatial experience despite the finesse of stereoscopically conjured virtual reality.

We produced a series of accurately rendered interiors taking into account materials, position of the sun, and latitude, the results of which are shown in Figs. 13, 14, 15 and 16.

Figs. 13, 14, 15 and 16 Rendered views of the interior of the Sala Creuer

I very much doubt that they yield any information that prepares the viewer with a presage of what they might experience in reality. Such commentary is somewhat anecdotal, of course, and at some future date when the Sala Creuer is completed it might be interesting to investigate this apparent gap between the familiar and unfamiliar more fully. Here I am claiming from the viewpoint of the designers, advances in digital prototyping offered significant evaluation of the relative advantages of different spatial configurations but in no sense do we hold any belief that we had been given any substantial insights into the likely spatial experience. If almost total cognitive unfamiliarity of a space is one reason for the failure of a prototype less than full scale to offer a proxy for the experiential quality of spaces perhaps it is scale itself that reduces the role of design prototypes in this regard; this is to say that the prototype is distinct from the mock-up, which necessarily needs to be 1:1 with a simulation of the materials at least, if not those actually proposed.

7 Learning from the Passion Façade Narthex

In this rethinking of prototypes in our post digital era perhaps we were never pursuing veracity per se but delving into new roles for prototyping when designing architectural elements and spaces, which in their own way, seem to be highly innovative. One of the smallest elements for the Sala Creuer was the hopper that collects all the water shed from the roofs and directs it down the drainpipe. This relatively minor object makes an important contribution to the external spatial configuration where roofs surfaces interface with walls and windows so was never quite so trivial as its role might suggest. Working with a parametrically flexible model gave us the opportunity to refine the design much more efficiently than would have been the case with explicit digital modelling. The actual workflow was very revealing about the problem of scale.

At least 20 significantly different versions of the rainwater hopper were produced, in a chain of development that could be radically different between versions, but nevertheless was highly iterative. Initial evaluation was based on the renders made along the workflow. At each significant shift in direction, a 1:25 3D print was made, always revealing room for improvement that had not been picked-up in the renders. Once we had a version that appeared to be viable at 1:25 we then modelled at 1:10. At 1:10 we saw room for improvement that had not been visible at 1:25.

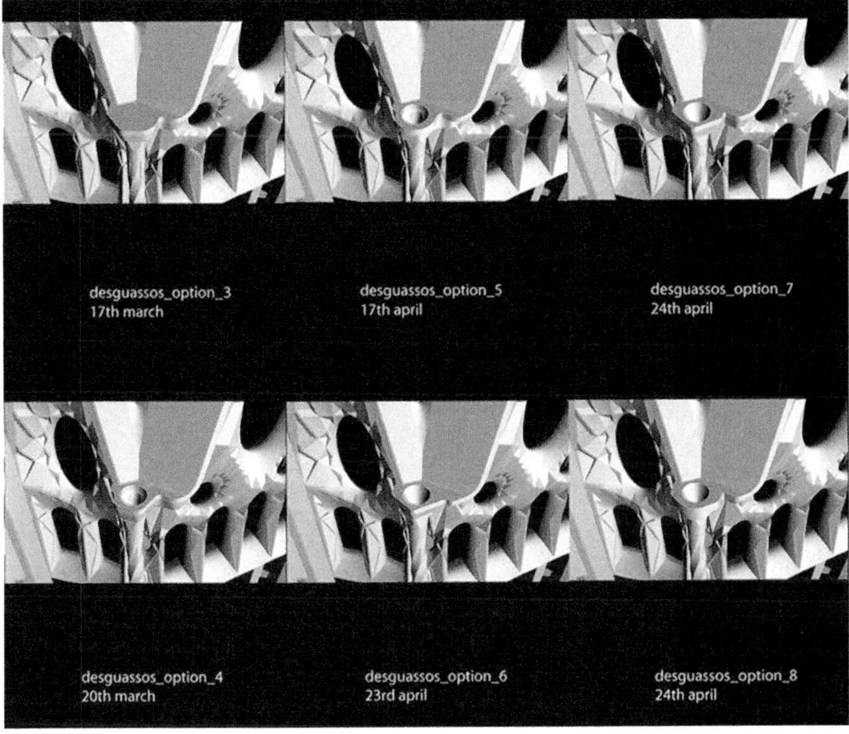

Fig. 17 Iterations in the development of the external Sala Creuer rainwater hopper

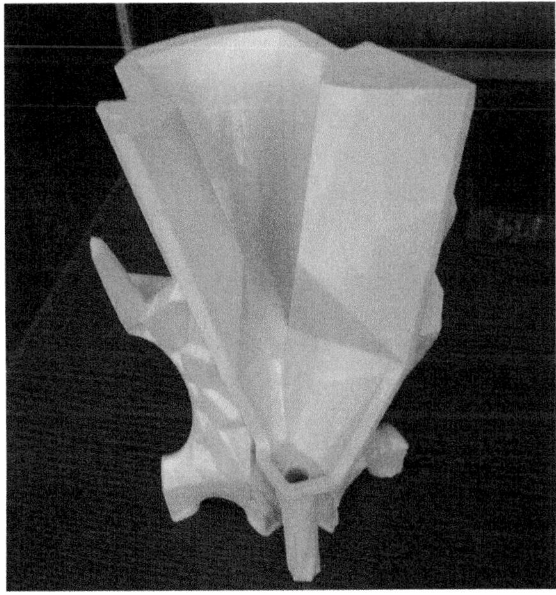

Fig. 18 1:50 model of the Sala Creuer rainwater conduit

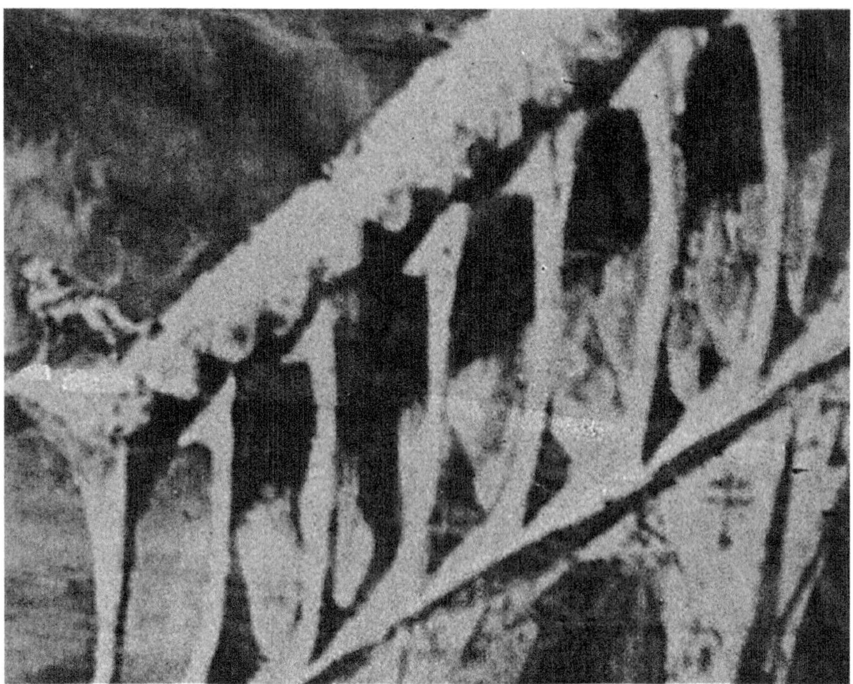

Fig. 19 Detail from the photograph of Gaudí's original drawing for his design of the Passion Façade narthex (Fig. 3)

Only when we had a version that worked at 1:10 did we commit the design to stone, yet to be revealed as built work at the time of writing (as it is still behind scaffolding) (Figs. 17, 18).

At the Sagrada Família Basilica we have used full-scale mock-ups taking on the role of prototype in the sense that they were tests of an approach rather than precise facsimiles of the intended outcome. One such example has been the column design for the narthex of the Passion Façade—definitive design commenced in 2001 with completion scheduled for late 2015.[2] The narthex sketch design had been undertaken for several decades before commencing the definitive design working not from any surviving plaster models but from a fine-grained photograph of Gaudí's fastidiously executed drawing of the façade (Fig. 19).

The combination of the inherent spatial complexity of Gaudí's design and the need to work within Gaudí's spatial palate of intersected hyperbolic paraboloids and hyperboloids of revolution (of one sheet) had proved to be too exacting a task for the various designers labouring by hand before the opportunities that 3D flexible modelling, 3D printing, and ultimately file-to-factory off-site fabrication offer

[2]The digital workflow for this part of the Sagrada Família Basilica project is covered in more detail in Burry (2011). *Scripting Cultures.* Wiley, London.

Fig. 20 Full-scale prototypes that could be moved backwards and forwards to gauge the optimum overhang

became possible. Given the prior difficulties faced by colleagues there was a lot at stake in bringing this design to fruition given the bold geometrical gymnastics involved. For that reason the team decided to make 3 columns at full-scale and place them in the actual position on site in the middle of the 9 column colonnade that forms each side of the overall composition. They were file-to-factory productions made from NC sculpted expanded polystyrene with an extraordinarily accurate painted finish to resemble the granite from which the actual columns were to be made. These were hoisted in position above the lower part of the façade, which had been completed in the late 1970s, and they remained there for several years. The hexagonal prism elements that are supported by the columns were also made (out of painted plywood), and were themselves parametrically variable as full-scale physical prototypes that could be moved backwards and forwards to gauge the optimum overhang in order to match Gaudí's drawing (Fig. 20).

Essentially mock-ups of the final columns they were prototypes working towards the definitive design. Needless to say, as the narthex colonnade nears completion, it is evident that these mock-up prototypes were not sufficient in themselves to presage the overall spatial experience (Figs. 21, 22, 23, 24 and 25).

Fig. 21 Expanded polystyrene mock-up

Fig. 22 Mock-ups in place acting as prototypes for future decision-making

Fig. 23 Final nine metre high column at the site of its NC cutting

Fig. 24 Narthex colonnade under construction (2014)

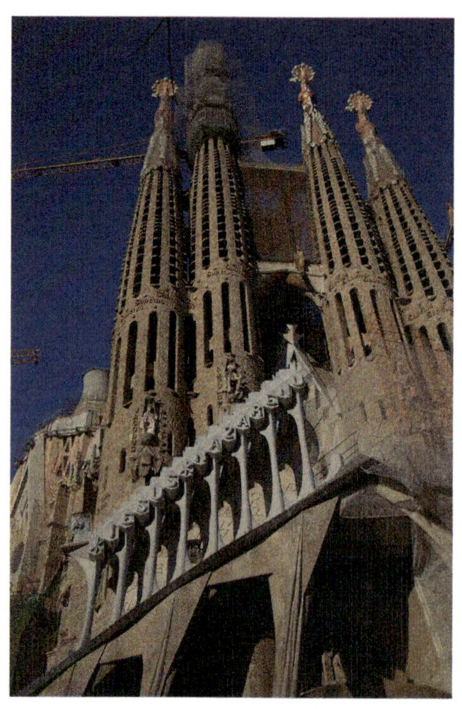

Fig. 25 Narthex colonnade completed on the left hand side (2015)

8 Prototyping Innovation in Offsite Fabrication

Returning to the Sala Creuer the stand-out prototyping in terms of innovation has been in the arena of digitally enabled off-site fabrication. The construction of this major conjunction of 5 tower bases and interior auditorium has demanded significant applied research in the ways and means to conduct the major building activities at ground level where possible. This has focussed on two areas of major game-changing: production of large-scale steel formwork offsite and the creation of permanent formwork using stone as masonry elements put together at ground level and hoisted into position. Typically the permanent formwork is the final exterior 'skin' of the building with the steel formwork defining the interior concrete surface formulated as artificial stone and subsequently bush-hammered to match the masonry. The steel formwork is made from elements laser-cut to millimetre precision and the off-site assembled masonry works with similar tolerances. Constructed at full-scale the Sala Creuer space itself may or may not be a prototype for spaces in a similar context—most unlikely, but its role as prototyping advances in construction for complex architecture seems to be beyond question (Figs. 26, 27 and 28).

Fig. 26 Sala Creuer under construction (2010)

Fig. 27 Detail of internal formwork and external permanent formwork

Fig. 28 Some hints of the eventual spatial experience noting that in this image neither the elaborate ceiling vaults nor the upperside of the ceiling vaults to the crossing below are yet visible

9 Concluding Observations

Completing the Sagrada Família Basilica is a singular challenge yet it has consistently offered windows to future architectural endeavour. Just as its early incursions into flexible (parametric) modelling in the early 1990s accidentally pointed to what have become mainstream design approaches 20 years later we might presume that pushing the design of the Sala Creuer and Passion Façade narthex offer insights to what are most likely rapid shifts to more comprehensive offsite fabrication. The signs are already there. The principal take-home message from the design processes involved for the Sala Creuer space is that we cannot successfully prototype spatial experience when the subject of the prototyping is unfamiliar in terms of its spatial attributes. Perhaps more experientially accurate stereo 3D virtual reality or hologram science advances are just around the corner, but at the time of writing it seems that for projects of this nature we have to await the completed building before we can fully appreciate spatial qualities of design. 3D printing of scaled models of design pathways represent highly significant advances in the role of physical prototyping architectural configurations albeit at an abstract level that could only have been dreamed of a quarter of a century earlier, still less in Gaudí's time. When considering what Gaudí achieved in his day for setting-out a schema with which to complete the Sagrada Família Basilica using what were then relatively limited resources points us to the ascendency of human conceptual ability and spatial understanding relative to any advances in prototyping to date. It serves to remind us that prototyping is still a rapidly evolving a means to an end and might best not presumed to be the a priori seeds of creative endeavour.

Acknowledgments The design team for the Sala Creuer consisted of Jordi Bonet, Jordi Faulí, Jordi Coll, Marta Miralpeix in Barcelona, and Mark Burry, Barnaby Bennett, Daniel Davis, and Michael Wilson in Melbourne. The design team for the Passion Façade narthex consisted of Jordi Bonet and Xisco Llabrés in Barcelona, and Mark Burry and Jane Burry in Melbourne. Much of the design research reported here was supported by the Australian Research Council. This additional support has been within the framework of a longstanding commitment by the Sagrada Família Basilica Foundation to commission university design research teams to assist investigations on site into innovative design, design representation, and digital fabrication.

Reference

Burry, M. C. (2011). *Scripting cultures*. London: Wiley.

Part II
Rethinking Prototyping

The Evolution from Hybrid to Blended to Beyond Prototyping

Kai Lindow and André Sternitzke

The traditional understanding of prototyping among different disciplines comprises technological and conceptual limits. With respect to user-oriented design of complex products, systems and services, new opportunities are emerging through innovative information, communication and manufacturing technologies. The growing technical complexity and the increasing individualization of products in turn require intelligently designed representations and test environments. In this way, design, production and interaction processes can be optimized for the respective users.

Research in this field requires the collaborative investigation of engineering and creative design disciplines covering close-to-engineering prototyping, the integration of mobile communication into prototyping and alternative design and production processes beyond prototyping.

Three mixed research groups from this research institution, along with two universities, Technische Universität Berlin (TU Berlin) and the Berlin University of the Arts (UdK Berlin), committed to work together in a new hybrid form by applying their complementary research expertise in order to investigate different prototyping perspectives in a symbiotic approach. Contemporary concepts and alternative models infuse the traditional and creative development processes by means of new prototyping elements.

Within the project, the diversity of ideas that are associated with different methodologies and discipline-specific approaches were combined in order to create a new transdisciplinary understanding of prototypes and prototyping. This approach necessitated the transdisciplinary cooperation of the involved disciplines because the issue goes beyond a single professional or disciplinary definition. The integration of different disciplinary perspectives, the creative design and the applied

K. Lindow (✉)
Industrial Information Technology, Technische Universität Berlin, Berlin, Germany
e-mail: kai.lindow@tu-berlin.de

A. Sternitzke
Institute of Architecture and Urban Planning (IAS), Berlin University of the Arts, Berlin, Germany
e-mail: sternitzke@udk-berlin.de

C. Gengnagel et al. (eds.), *Rethink! Prototyping*,
DOI 10.1007/978-3-319-24439-6_7

Fig. 1 Overview of the "Rethinking Prototyping" research project

engineering perspectives serve to comprise the transdisciplinary approach (cf. Part III in this volume). In this way, scientific investigations are defined from different design perspectives. Due to the fact that the transdisciplinary working principle goes beyond modest networking or solely linking multiple disciplines, this flexible approach grants participants the opportunity to reach out to the core of understanding the concept and the operation of prototypes and prototyping. The transdisciplinary research groups' knowledge and methods were interlinked and integrated in order to raise awareness of different prototyping definitions and to investigate the prototyping aspect of various scientific working principles and competencies (Fig. 1).

The research partners involved developed a common understanding of prototyping in an iterative process. Moreover, the interfaces of a transdisciplinary design process were examined and the divergences of prototypes for productive design approaches were investigated. The following chapters offer novel insights and findings about the hybrid, blended and beyond prototyping approaches. As part of the transdisciplinary research project the prototyping streams were addressed as below.

The research stream "Hybrid Prototyping—New approaches of prototyping for testing and validation of integrated products and services in the context of urban living space" investigated the role of prototyping for the integrated development of products and services, so-called Product-Service Systems (PSS). The development of PSS prototypes was placed in the context of rapidly changing urban environments in order to explore new utilization concepts, use of thresholds and related design options. On the transdisciplinary basis of this research stream, user needs had been identified in the urban living space and were investigated in representative scenarios. The hybrid combination of products and services was tackled by means of a hybrid prototyping combination of physical prototypes and digital models in virtual reality (VR), thus linking two different prototyping fidelities for enabling a PSS realistic experience. The research stream "Blended Prototyping—Research and development of mixed prototypes for mobile communication" linked research of design and styling in engineering with research of usability in software engineering. Based on the new "Blended Prototyping" approach, it demonstrated that the benefits of low-fidelity prototyping are retained while a sufficiently product-driven interface can be provided in order to extend the coverage area of usability problems. This project combines different prototyping fidelities as well, yet takes another step and merges them in such a way that boundaries between the prototyping approaches disappear. The research stream "Beyond Prototyping—Opportunities and limitations of alternative design and production processes beyond prototypes" investigated the role of prototyping in the focus of the novel production technology "Rapid Manufacturing". It addressed specific design and technological issues, as well as economic ones, and also focused on the role of stakeholders in the design process. The prototype will likely become obsolete while the customer self-designs each product which in and of itself can be regarded as a unique piece. Thus the meaning evolves from experiencing and usability testing to eventually become the final product itself. The following chapters describe these main research streams in detail.

Hybrid Prototyping

Konrad Exner, André Sternitzke, Simon Kind and Boris Beckmann-Dobrev

Abstract Innovative ideas and solutions are a decisive competitive advantage in today's global markets. The concept of Product-Service Systems (PSS) integrates services, products, infrastructure and business models in an individual solution for the customer. In order to receive the full benefit in providing PSS systematic development methodologies are needed to cope with the complex structure of these systems. Hybrid Prototyping combines physical prototypes and digital models in Virtual Reality. The utilization of this concept enables a prototyping of PSS in early development phases. The main objective is the integration of the customer in this process and enable a realistic experiencing of PSS concepts in order provide the means for the validation of PSS.

1 Introduction

For decades, engineering design processes have been changing and adapting due to constantly new trends and innovations. Researchers and companies are continually developing, evaluating and integrating new methodologies to optimize their processes, increase quality and reduce costs. A phenomenon of note in many urban areas is a declining demand by middle-class customers to own goods as a status

K. Exner (✉) · S. Kind · B. Beckmann-Dobrev
Industrial Information Technology, Technische Universität Berlin, Berlin, Germany
e-mail: konrad.exner@tu-berlin.de

S. Kind
e-mail: simon.kind@tu-berlin.de

B. Beckmann-Dobrev
e-mail: boris.l.beckmann-dobrev@tu-berlin.de

A. Sternitzke
Institute of Architecture and Urban Planning (IAS), Berlin University of the Arts, Berlin, Germany
e-mail: sternitzke@udk-berlin.de

C. Gengnagel et al. (eds.), *Rethink! Prototyping*,
DOI 10.1007/978-3-319-24439-6_8

symbol (Miller 2014). The reasons differ from sustainable to financial aspects or simply a changed consumer behavior. Product-Service Systems (PSS) offer solutions instead of products for customers in business-to-business as well as business-to-customer sector. The development and validation of these integrated products and services is of high relevance in research and industry regarding engineering design and economics.

1.1 Motivation and Objectives

Global and fast changing markets as well as transitioning urban areas create challenges for governments and companies. As of 2008, more than half of the world population is living in urban and mega-urban areas. Accordingly, new concepts regarding mobility, living and energy management need to be developed (Gengnagel 2011) in order to cope with increased pressure on the infrastructures of these cities. Furthermore, cultural developments in multicultural cities generate new trends and mentalities in society, resulting in changed mindsets regarding consumption and lifestyle. The increasing demand for new solutions and services requires holistic and systematic approaches in order to integrate these different areas and challenges. It is against this backdrop that the research team turned its focus to the concept of Product-Service Systems. PSS do not merely extend the product centered view in engineering design with added services; PSS are signified by the holistic and integrated development of a system including products and services (Aurich et al. 2004). As a result, the development of PSS requires transdisciplinary teams due to the inherently diverse system elements (Exner et al. 2014a). In order to deal with the broad project content, the researchers of Technische Universität Berlin (TU Berlin) and the Berlin University of the Arts (UdK Berlin) complement the specific knowledge and background of their disciplines to create new approaches and solutions.

The main idea of PSS is to offer customer solutions for fulfilling specific customer needs. For instance, car-sharing concepts provide the customer a mobility solution without the need of owning a product, namely, a car. Companies expect PSS to generate strong customer loyalty, increase customer value, achieve competitive advantages and raise the company's revenue with services (Mont 2002). Nevertheless, these potential company benefits can only be fully exploited by means of an integrated development process. For instance, car sharing has not been developed in an integrated manner, but added services and infrastructure to an existing product, thus reducing potential added values. Therefore, the basis of a successful PSS development and implementation is based in a systematic PSS development process (Shimomura et al. 2009). The customer is, as already indicated, another key factor, because a PSS should satisfy specific customer needs. For this reason, the integration of the customer in the idea generation process as well as in the validation process of first PSS concepts is an important factor (Mannweiler 2010).

The main objective of this research project is: To develop and evaluate new prototyping methodologies for PSS in early development stages and enable the integration of the customer in this process. Validation with prototypes is a common and widespread concept across all disciplines and is also the main focus for the development of PSS validation methodologies in this project. For this purpose, the model of Product-Service Systems has been applied in a use case for urban mobility. In summary, the main points of emphasis of this research project are:

- New mobility solutions in urban areas
- Prototyping methodologies for PSS
- Integration of the customer

Recent research regarding PSS development methodologies indicates a lack of validation methods. In particular, prototyping of PSS can be seen as a desideratum in research and practice. In order to fill the gap, new ways of prototyping for PSS have to be created, tested and evaluated. Regarding the complexity of PSS, a transdisciplinary approach seems most promising due to the combination of different perspectives and expert knowledge. Regarding PSS development methodologies and new solutions for urban mobility, the following research questions (RQ) have to be examined:

- [RQ1] How can the demands of customers be analyzed to generate ideas and solutions for new mobility concepts in urban areas?
- [RQ2] How can PSS be prototypically implemented in order to validate PSS in early development stages, like planning and concept phase?
- [RQ3] How can the integration of the customer in the validation process be realized?

In addition to the analysis of existing validation methods, a first step is the adaption of the Smart Hybrid Prototyping approach that has been developed by Beckmann-Dobrev et al. (2010) at the Chair of Industrial Information Technology of TU Berlin. The main idea is the validation of mechatronic systems in combining physical prototypes and virtual models in a Virtual Reality. Both Product-Service Systems and mechatronic systems can be characterized with the complexity due to interdependent system elements. Regarding mechatronic systems, the integration of mechanical aspects of the product with informatics and electronics has already been achieved with SHP. Therefore, complex interactions have been enabled and can be enhanced with additional element, like services. For this reason, the feasibility of using this approach for prototyping of PSS seems promising. Regarding this topic, the following research question can be formulated:

- [RQ4] How can the Smart Hybrid Prototyping approach be adapted in order to prototype PSS?

In summary, the objective of this research project is the development of a prototyping methodology for PSS and the integration of the customer in the PSS development process. Figure 1 shows a first vision of the new approach. It symbolizes the envisaged interaction of the user with the product in Virtual Reality.

Fig. 1 Pedelec and user in virtual reality (Exner and Stark 2015)

1.2 Structure of This Chapter

The "Hybrid Prototyping" subproject integrates different points of view regarding the disciplines and research topics. In Sect. 2, the perspectives on hybrid prototyping as well as the connections between them are introduced. Section 3 describes the overall research approach and procedure for developing the methods. In Sect. 4, the analysis regarding the state of the art and important validation methods in respect to their relevance for hybrid prototyping is conducted. In Sect. 5, the actual findings, meaning the methods and results of the evaluation can be found. Section 6 concludes with a critical discussion of the results and an outlook for future research.

2 The Hybrid Prototyping Perspective

The term "hybrid" derives from Greek and means something bundled, crossed or mixed. The manifestation of "hybrid" in the "Hybrid Prototyping" subproject has to be considered with respect to three perspectives. One the one hand, Product-Service Systems which combine tangible products and intangible services to an integrated systems, thus providing specific solutions for the customer (Sakao and Lindahl 2009). In this case *hybrid* refers to product and service development methodologies and the research focus is on integrating the development and the validation of both. On the other hand, the technical implementation of a new prototyping concept to enable the validation of PSS with a new prototyping approach determines the second perspective. The Smart Hybrid Prototyping (SHP) approach combines physical prototypes and digital models in a Virtual Reality in order to enable a realistic experiencing of a mechatronic system (Beckmann-Dobrev et al. 2010; Stark et al. 2009). The *hybrid* aspect is the combination of contrasting physical and virtual elements in order to enable prototyping of PSS, which means the simultaneous testing of product and services. Finally, in architecture all interferences that affect objects, environment and infrastructures can be designated as "hybrid" due to the architectural intervention in the environment, as well as in the infrastructure. An important aspect is the interaction of architecture with mobility infrastructure and the consideration of the consequences (Pinto de Freitas 2011). Furthermore, the creative design aspects of architecture enrich the PSS and SHP emphasis with new insights due to a diverse perspective of prototyping.

2.1 Product-Service Systems in a Prototyping Perspective

PSS represents a system of products, services, infrastructure, software, provider network, etc., in order to provide customer specific solutions. The novelty of the concept is the systematic approach and the integrated development of the PSS to ensure the consideration of the interdependencies between system elements (Meier and Uhlmann 2012). Therefore, the PSS concept is much more comprehensive than value-added services to existing products. Due to the complexity of the system with related elements and various network partners, existing development methodologies are insufficient for PSS development processes. In order to fill this gap, research has been ongoing and PSS specific methodologies have been developed and evaluated (Shimomura and Arai 2009; Lindahl et al. 2006; Sakao and Shimomura 2007; Sadek 2009; van Halen 2005; Tan 2010; Matzen 2009; Müller 2014). In spite of this research, the focus is mainly on design methodologies, sustainability, business models and cost calculation. The validation of PSS lacks consistent and holistic approaches to ensure a systematic testing and evaluation of PSS concepts or system properties (Müller 2014). Furthermore, research regarding PSS prototyping can be considered a desideratum in academia and practice (Exner et al. 2013). For this reason, prototyping approaches of product, service and system development methodologies have been analyzed in order to develop a comprise understanding for PSS prototyping. A comprehensive analysis of validation methods in these areas is presented in Sect. 4.

On the one hand, prototyping in product development focuses on the validation at milestones and design reviews (Ulrich and Eppinger 2012). Therefore, physical product prototypes comprise the actual state of development and enable an evaluation for management and customer. One the other hand, digital models and simulation are used to verify product properties during the product development process (Pahl et al. 2007). In summary, tangible and intangible aspects are represented with physical and digital prototypes in product development design processes with an increasing tendency regarding virtualization of both. Prototyping in service development refers mainly to process prototyping, thus visualizing service procedures (Bruhn 2006). Besides, two more approaches in service engineering can be stated. Firstly, concept prototyping (Schmid 2005) by the use of storytelling and role-playing with focus groups. Secondly, the simulation of the service is another possibility by integrating the environment and necessary elements for a prototype of the service, thus enable experiencing of the service in a realistic environment. Due to the high effort, the research community started substituting the elements and environment with virtual models in Virtual Reality (van Husen and Meiren 2008). System development, better known as systems engineering, is focusing on systems that consist of interrelated components (Kossiakoff et al. 2011). In this way, it is a comparable development discipline to PSS, but lacks the emphasis on services and provider networks. Besides the classical product prototypes, simulations and thus prototyping of complete systems is a major aspect in systems engineering. Therefore, dynamic (time-dependent) and static simulation types have a high significance in prototyping of systems (Engel 2010). In conclusion, the classical disciplines often distinguish between low and high fidelity approaches, with a strong tendency towards virtualization.

In PSS development methodologies, prototyping has been rarely stated at all and validation methods do not refer to PSS prototyping. Nguyen et al. (2014) integrated prototyping in a PSS development process, but do not analysis the usage in different development stages and do not offer any methods or tools. The PSS-Inspector (Suvarna et al. 2010) integrated digital CAD models and process flow charts in order to enable an evaluation at design reviews and, thus can be seen as a prototype for PSS. Further discussion can be found in Exner and Stark (2015) and Exner et al. (2013). In summary, a consistent PSS prototyping does not exist, thus new PSS prototypes with an integration in existing PSS development processes are needed. The characteristics of a PSS-prototype can be defined as follow:

A PSS-prototype integrates tangible and intangible elements of the entire system in a single prototype. Due to the considerably differences of the PSS elements, yet high interdependencies, new hybrid prototyping approaches are essential. These have to integrate diverse aspects, like physical and virtual, in order to facilitate the complex interaction between elements of a PSS. Therefore, explorative prototypes enable an externalization of first ideas and concepts in early PSS development phases, like planning and concept phase, to discuss and reflect mental pictures. Evolutionary prototypes are intended to visualize and validate PSS solutions or intermediate results along the PSS design process. Furthermore, experimental prototypes can be used at all development stages to test properties of the PSS.

With respect to the development of an integrated PSS-prototype, one of the biggest challenges for this research team is the conflation of the different product and service characteristics. Firstly, services have a strong procedural, thus non-static character.

Furthermore, the simultaneously generation and consumption as well as the involvement of the customer in the process are main attributes of services. Secondly, products can be characterized due to their design, meaning aesthetical aspects and their inherent properties that can be operationalized. Today's customer demands a solution that integrates both perspectives, thus a PSS-prototype has to consider:

- Procedural process perspective
- Integration of the customer
- Aesthetical design
- Inherent and measurable properties

In order to develop such a complex PSS-prototype, the research team focuses on new digital and virtual approaches and techniques, namely the Smart Hybrid Prototyping, which is introduced in the next section.

2.2 *Smart Hybrid Prototyping in a Prototyping Perspective*

The main goal of Smart Hybrid Prototyping (SHP) is the support of interdisciplinary development of mechatronics products and subsystems combining prototyping technologies from physical and digital prototyping. In order to understand the idea behind and the need of SHP, an overview of some other technologies building the base of SHP is required. In particular, these are visualization technologies like Virtual, Augmented and Mixed Reality, modeling and simulation technologies with physic based game engines or professional tools like MATLAB/Simulink or Dymola, interaction technologies and human machine interfaces (HMIs) like computer haptics and Tangible User Interfaces (TUIs).

Virtual Reality (VR) is a high-end visualization technology. According to Burdea and Coiffet (2003), VR should be interactive, immersive and encourage the human imagination. Especially interactivity and immersivity make the difference between VR and 3D cinema. The continuum between Virtual, Mixed and Augmented Reality (VR, MR, AR) defined by Milgram et al. (1995) is a fluent transition from virtuality into reality. Milgram describes it as a mix of virtual and real portions where the relation between physicality and virtuality defines a fluent stage between VR with predominant virtuality, or AR with predominant reality. The continuum between VR and AR is described as MR (Milgram et al. 1995).

The main application of VR in industry is carrying out design reviews (DR). DR's are important milestones within the product creation process in order to ensure requirements are fulfilled, the quality, to solve issues and to make decisions. To prepare the design review, a Digital Mockup (DMU) from the CAD or PDM system must be derived and imported into the VR environment. A DMU is a static representation of the actual development stage of a product (Spur and Krause 1997; Krause et al. 2007a).

In order to extend the virtual product up to functional behaviors, a Functional Mockup (FMU) was described that is characterized by functionalities that are typical for mechanical systems like kinematic and dynamic constraints which are mostly animated or allow limited interaction with the DMU using standard input devices, such as a mouse and keyboard (Krause et al. 2007b). One limitation of FMUs is the ability to represent functional behaviors of mechatronic systems. Therefore, a Functional Digital Mockup (FDMU) was defined (Stork et al. 2010). FDMUs are able to represent the functional behavior of mechatronics systems by extending mechanical functions of FMUs with logical state transitions driven by sensoric and actoric events. FDMUs are suitable for Software in the Loop (SiL) validation or Hardware in the Loop (HiL) verification of mechatronics products in the early development stage. In practice the Functional Mockup Interface (FMI) is implementing the FDMU concept partially by providing standardized interface for co-simulation and data exchange. The support from the tool vendors to implement the standard in their tools is required (Schneider et al. 2009).

Smart Hybrid Prototyping (SHP) is a modern prototyping technology. It expands the validation and verification approach of the FDMU method slightly to a Mixed Reality continuum similar to the Milgram continuum. The main goal of SHP is the support of interdisciplinarity within prototyping of mechatronics products and subsystems. SHP covers the overall product creation process, from the early stage up to start of production. The idea is to create one digital prototype at the beginning of a project that grows fluently during the development stages from a virtual to a physical prototype. The continuum in between is defined as a hybrid prototype and similar to MR depends the relation of physical and digital parts on the stage within development process. The smartness of hybrid prototyping arises from its interdisciplinarity. The combination of many technologies from mechanical engineering, software, electrics and electronics as well requires deeper knowledge of modern development methods and the subset of corresponding technologies. The ability to combine them into a hybrid functional prototype cannot be ensured by individuals, but rather only by interdisciplinary teams. Therefore, SHP encourages interdisciplinary collaboration. SHP technology does not need an adaption of the common development process and can be added to proven milestones, such as design reviews.

2.3 *Prototypes in Architecture*

The term "prototyping" has only appeared in the architectural world in recent times, as a result of new technologies that are being used. The development of a 1:1 prototype of an entire architectural design is not yet possible. Each resulting building is a prototype in and of itself. When describing the scale depiction of designs, the term model is used; both for real (built) objects as well as for virtual (digital) representations. There are a large variety of different types of architectural model, which are used for various different types of scenario. The most frequently used types of model are:

- The working model: A model that is mostly made from easily workable and inexpensive materials which serves to test designs in early development phases. The most important aspect is not precision, but the ability to quickly create spatial objects that are easy to modify and adapt. Therefore, the working model can be described as a spatial sketch.
- The design model: Serves as a representation of the design in the intermediate stage of the design process and establishes the basis for discussions between developer and planner.
- Competitive model: A model of the architectural design created for competitive purposes.
- Presentation model: A carefully prepared model, for presenting the design to developers or the public. It is mostly complex, true to detail and made from high quality materials. It presents the architectural design accessible in a way that is also understandable to laypeople.
- Urban model/Environmental model: Sets the design within the context of its immediate environment. This could either be an urban or suburban context.
- Solid model: Depiction of the structure using simple solid bodies only. It is mostly used for small-scale designs.

Nevertheless, prototypes prove more important for the architectural design. They allow the review of individual parts of the building (i.e. facade 1:1 prototype) as well as the later use (i.e. climate sustainability). The term prototyping is used in three different ways. Firstly, it utilizes new rapid prototyping technologies that enable an easy creation of models whilst permitting new levels of accuracy and the use of new materials. For instance, the CNC laser cutting of thin panel material and board benefits of this technology. This process permits a level of accuracy and detail that just would not have been possible using tools such as cutters (knives) and scissors. For manufacturing mass models, 3D CNC milling or 3D CNC printing is used instead. At first, all these technologies were used purely as a tool for creating models. Their high accuracy and speed enable rapid testing of different options. Their level of detail is higher than before.

Secondly, these technologies are no longer only used for depicting architecture, but also for creating points of detail, construction systems and even small pavilions in 1:1 scale. Even in early design stages these prototypes are used to verify and test construction principles. In this perspective, in the world of architecture, we speak of prototypes. They can be in physical form, such as milled or printed construction elements, but also in digital form. These digital prototypes like simulations or virtual spaces help planners to test and examine architectural, structural or climatic (see Fig. 2) concepts right in the building's very early design stage, and also enable developers to experience their planned buildings in Virtual Reality, using technologies such as head-mounted displays or Cave Automatic Virtual Environment (CAVE) (Meibodi and Aghaiemeybodi 2013).

Fig. 2 Visualization of the thermal performance of a specific design/digital pre-fabricated, custom folding models (Diploma Steffen Samberger)

The combination of digital and analogue technologies, in the meaning of hybrid, inherently possesses enormous potential for the future, and enables the development of new design strategies and technologies, right up to the ability to be able to validate design decisions (Drewello 2013). Based on the experience with using of hybrid prototypes at the interface between buildings and urban environments new needs arise, forming the basis for future Product-Service Systems. With the help of architectural prototypes before erecting buildings or urban spaces can be alerted to necessary Product-Service Systems. Smart Hybrid Prototyping enables the testing of PSS in an architectural context. The resulting insights could create new possibilities for the architectural design process. For instance, the integration of SHP and PSS enables a high predictability of the interaction between buildings and urban space.

3 Research Approach

Due to the transdisciplinary character of the research project with an emphasis on collisions and cross-fertilization of ideas, a systematic scientific approach was necessary to ensure the quality and progress of the projects objectives. Therefore, different studies have been planned and conducted in order to develop and evaluate the project results. The research team applied the Design Research Methodology (Blessing and Chakrabarti 2009) in order to ensure an accurate scientific working method. The approach consists of 5 phases (see Fig. 3).

In a first step, a descriptive study comprises a comprehensive analysis of existing validation methods and assess their value for PSS-prototyping. A particular focus has been on the Smart Hybrid Prototyping approach due to the high potential for adapting this concept for PSS-prototyping. Additionally, relevant perspectives and dimensions regarding PSS-prototyping have been determined, thus specifying the application area for PSS-prototyping. The next step focuses on the development of a use case in order to provide test scenarios for the new methods. Therefore, a

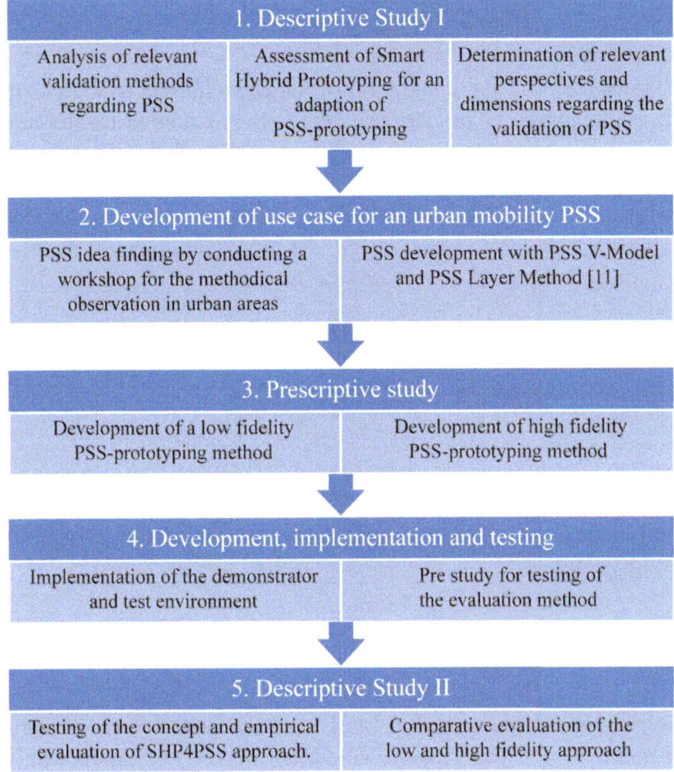

Fig. 3 Research approach

creative workshop identified the customer needs regarding urban mobility by applying the method cultural probes (Martin and Hanington 2012). The development of PSS concepts has been conducted with PSS V-Model and PSS Layer Method (Müller 2014) in order to support a systematic and integrated PSS development. Based on the results of the analysis in the first step and the use case for urban mobility two methods have been developed in a prescriptive study. On the one hand, a low fidelity method adapting the utility analysis for PSS (Exner et al. 2014c). One the other hand a method based on the Smart Hybrid Prototyping approach ensured a high fidelity perspective (Exner et al. 2014b). Both methods consist of different elements that needed to be developed in order to implement the methods and enable a first testing. The components differ greatly from evaluation matrices to physical interaction devices and digital models in a Virtual Reality. In a second descriptive study, a qualitative and quantitative evaluation ensures the feasibility of the methods and enables a comparison between the low and high fidelity approach.

4 Relation to Existing Theories and Work

PSS are characterized by the integration of various different system elements. For this reason, the analysis of validation methods is not limited to product development, but to service and system elements as well. Furthermore, the integration of different stakeholders and their perspectives need to be considered. Additionally, the Smart Hybrid Prototyping approach is analyzed and presented in order to assess the applicability for the prototyping of PSS. Section 4 summarizes the results of the Descriptive Study I (Fig. 3).

4.1 Validation Methods and Perspectives

In this section the different perspectives of developer, customer and manager regarding the validation process of PSS will be described. Furthermore, a synopsis of validation methods in relation to their discipline evaluates a possible utilization for PSS-prototyping.

4.1.1 Validation Methods

PSS-prototyping approaches need to integrate validation aspects of various disciplines and methods. For this reason, validation methods of product development, service development and PSS development have been analyzed for their relevance of PSS prototyping regarding the important validation dimensions: concept, customer and interaction (see Table 1).

Additionally, some methods regarding the validation of PSS have been presented in research, yet do not consider PSS prototyping. Further details and discussion of PSS validation methods can be found in Exner and Stark (2015) and Exner et al. (2014b).

4.1.2 Validation Perspectives

The validation of PSS needs to be considered regarding two questions. (1) Which validation dimensions for PSS exist? (2) Which perspectives in relation to these perspectives need to be reflected? Burger and Schulz (2014) conducted a study with technical services providers. As a result of the study, ten important dimensions have been derived. Stark et al. (2009) specified three important perspectives for the validation process of mechatronic systems: customer, developer and decider. Exner and Stark (2015) transferred and evaluated these aspects in order to assess their relevance for PSS-prototyping, see Table 2.

Table 1 Validation methods

	Validation methods	Valuation for PSS prototyping
Product development	Design Reviews: sketches, 3D draws and models, physical prototypes	The form of design reviews provides important ideas regarding the representation of ideas concepts
	Mock up: Physical Mock Up (PMU), Digital Mock Up (DMU), Functional Mock Up (FMU)	Mock ups are an essential factor for PSS-prototyping in order to assess the product elements of the PSS
	Computer-Aided Engineering (CAE): N-body simulation, finite element method (FEM)	CAE is very product centered approach in later development phases, thus not important for early PSS prototyping
	Simulation with Virtual Reality (VR)	Simulations in VR could enable the developer to represent the complex correlations between PSS elements
	Smart Hybrid Prototyping (SHP)	SHP enables the validation of complex mechatronic system elements; hence PSS prototyping should be feasible
	VDI 2225	VDI considers both technical and economic components, thus could enable an integrated validation of PSS
Service development	Service Blueprinting	Service blueprinting is a pure process analysis with flow charts. It divides the process through the line of interaction, the line of visibility and the line of internal interaction. Based on this visualization service processes can be structured and optimized
	Simulation (physical): service-theater, service script, simulation (digital) with VR	The interaction with the service environment and customer should be integrated in PSS prototyping. The substitution of the environment with digital models complies with SHP
	Quality Function Deployment (QFD)	QFD considers the process, but does not enable interaction with the customer regarding concept ideas
	Failure Mode and Effect Analysis (FMEA)	The FMEA support an analytical approach, but not the chosen dimension for PSS prototyping
	Design Structure Matrix (DSM)	The DSM support complex systems as well activities, thus should be considered for the validation of PSS variants

(continued)

Table 1 (continued)

	Validation methods	Valuation for PSS prototyping
Systems engineering	Simulation: dynamic simulation, static simulation	Static simulation contradicts the procedural character of PSS, but the dynamic simulation of a system corresponds with the SHP approach and prototyping of the process
	Agent based systems	Agent based systems enable the validation of system behavior, but does not focus interaction and integration with the customer
	Utility Analysis (UA)	The UA enables the evaluation of different variants for several criteria. The validation of complex system concepts is not possible in the classical form, but the ease of use is positive

Table 2 Relevance of validation dimensions for perspectives of PSS (Exner and Stark 2015)

Dimensions	Perspectives[a]		
	Customer	Developer	Decider
1. Process	◕	●	◑
2. Concept	◕	◕	●
3. Resources technology	◔	◕	◑
4. Resources employee	○	◕	◕
5. Contact to customers	◑	◑	◑
6. Customer acceptance	◕	◑	●
7. Interaction	◕	●	◑
8. Customer reaction and emotion	◑	◑	●
9. Technical requirements	◑	●	◑
10. Variables service environment	◑	◕	◑

[a]Nomenclature
○ No importance
◔ Minor importance
◑ Medium importance
◕ High importance
● Very high importance

Four critical dimensions regarding the perspectives can be stated as a qualitative result of the analysis: process (1), concept (2), customer acceptance (6) and interaction (7). In order to comply with these constraints a new PSS-prototyping approach should focus on PSS concepts, enable interaction with the PSS, include the process character of services and integrate the customer.

4.2 *Smart Hybrid Prototyping*

This section describes visualization techniques used by Smart Hybrid Prototyping technology and shows some application examples. Furthermore, different advantages using SHP technology are presented (interaction, immersion, fidelity level). The relation to the validation perspectives is discussed regarding the integration and adaption of SHP technology for PSS development.

As illustrated in Sect. 2, Smart Hybrid Prototyping technology is using different visualization techniques used also by Virtual Reality (VR), Mixed Reality (MR) and Augmented Reality (AR). Function-oriented technologies like FMU and FDMU are used for realistic impressions. Several visualization technologies are suitable in the context of passing milestones during product development processes by the use of Design Reviews. Basically, SHP can be described as an evolution of Design Reviews, because the technology can support developers in demonstrating their work progress to managers and decision makers. High-end visualization technologies like CAVEs (Cave Automatic Virtual Environment) and Powerwalls, but also technologies from the gaming and entertainment sector can be considered, e.g. Oculus Rift. Different advantages can be taken into account regarding level of immersion, spatial visualization and also mobility, high level of maturity and low price of entertainment electronics. Dependent on the specific requirements and use cases in which Smart Hybrid Prototyping is used the most appropriate visualization technology can be chosen regarding the level of digital virtuality or physical reality needed, e.g. see Fig. 4.

Another important aspect of Smart Hybrid Prototyping technology is the ability of user interaction. By providing the ability for users to interact with digital models and virtual environments, the degree of immersion and perception can be particularly increased (Stark et al. 2010). A high level of immersion is desirable because impressions of realness and authenticity are increased by the quality and number of human senses involved. This is realized by extending the visualization technology by physical, mechatronic interaction devices that cover the needed spatial degree of freedom. Those physical components are linked and integrated into the visualization environment and perform a physical behavior. By the combination of virtual and physical components it is then possible to increase the level of detail on both

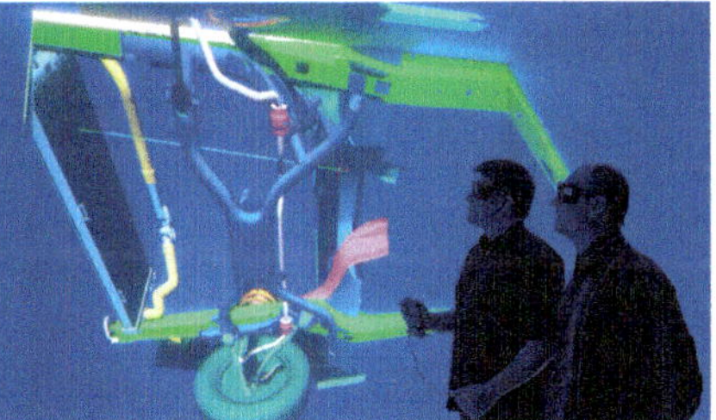

Fig. 4 Visualization techniques used by smart hybrid prototyping (Stark et al. 2009)

sides (virtually and physically) and allow earlier adjustments and optimizations during the product development process. The application of Smart Hybrid Prototyping technology also enables the potential of an early evaluation of products. The integration of customers, especially, is very attractive for an early feedback during product development process. Further advantages and a more detailed description of Smart Hybrid Prototyping technology can be found at Stark et al. (2009).

An application of Smart Hybrid Prototyping technology is provided in Fig. 4. By moving the physical component of the SHP device, the virtual model of the car tailgate moves in the same physical behavior as a real car tailgate. Haptic feedback can also be included through the integration of a simulation model of the tailgate of a real car. This application demonstrates the potential for an early product evaluation at the end of design phase for digital product models compared to functional testing with cost intensive physical mock-ups, which are available much later during product development process (Auricht et al. 2012). Other applications can be realized to evaluate the physical behavior of opening and closing a car door or to evaluate manual handling tasks during production phase. In this case, the weight and movement of real goods and products can be simulated within virtual environment for the use case of a manual assembly line.

The advantages of Smart Hybrid Prototyping technology are the degree of user interaction with digital models and virtual environment and the potential of the evaluation with early customer feedback to test acceptance of the future product. With regards to prototyping processes, the ability of increasing or lowering virtual or physical fidelity of prototypes also possesses a valuable potential of easily building and adapting different product solutions and approaches at lower costs. During concept phase, especially, this ability allows for an improvement of product design and functionality.

This leads to the challenge of adapting Smart Hybrid Prototyping technology for prototyping of Product-Service Systems. During development phase of PSS, basically the same requirements (concept, customer acceptance and interaction) are required for developing successful PSS. Due to the fact that intangible Product-Service Systems processes are difficult to visualize interactively, this challenge needs to be addressed by an appropriate adaption of Smart Hybrid Prototyping for Product-Service Systems (SHP4PSS).

5 Findings

According to the research approach presented in Sect. 3, four steps have to follow the initial analysis and results of Sect. 4. In order to apply and test new methods, an exemplary PSS development for an urban mobility PSS has been conducted. Based on this use case, two methods have been developed to enable the validation of a PSS. Firstly, a low fidelity approach based on the utility analysis provides an elementary validation of PSS. Secondly, a high fidelity concept with Smart Hybrid Prototyping enables the experiencing and actual Prototyping of PSS. Afterwards, the implementation of the new approach is described. The evaluation with a case study presents the results regarding the feasibility and the comparison of the new methods.

5.1 PSS Use Case for Urban Mobility

In order to test the new methods, a use case is essential for the evaluation of the research results. This particular use case should address the social questions, along with the resulting needs, arising from the urban space. The main task of the urban space is satisfying the needs of the people in the urban environment: living, working and recreation. From these arise the three most important principle needs of urban spaces: energy, communication and mobility (Baum 2008).

When we look at future population growth, along with the trend towards urban living, new demographical, ecological and social challenges arise. These have an effect on our economic and urban life, especially where our mobility is concerned. In future, urban mobility will involve less energy, lower costs, fewer effects on the environment and climate, less noise and lower levels of exhaust fumes and greenhouse gases (Hall et al. 2000). These objectives can only be accomplished by means of new, user-orientated and sustainable mobility concepts that will allow independent, flexible transport. For this reason, the research team conducted a case study in order to develop such a mobility concept for the future with the assistance of a PSS development methodology. The main objective was validating PSS concepts and creating the basis for later SHP test scenarios. In a first step, the customer, market and environment need to be analyzed in order to support the idea generation (see Fig. 5).

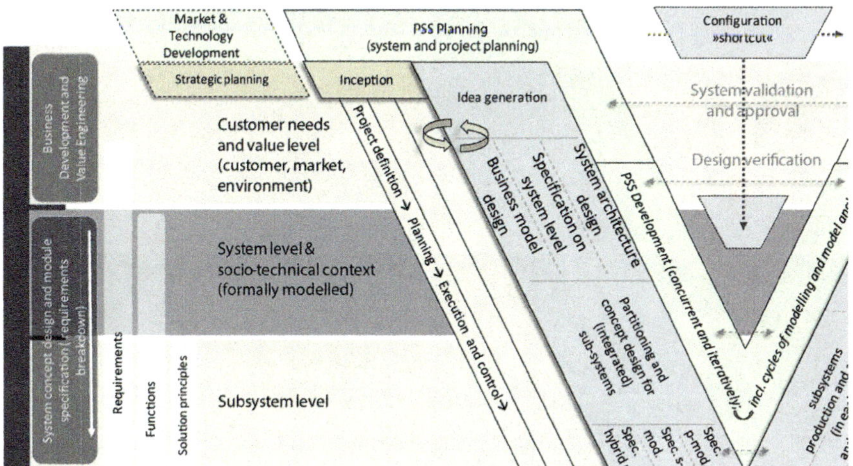

Fig. 5 Segment of the generic PSS development process model: "PSS V-Model v2012" (Müller 2014)

The target of this study was older people. This section of the population is steadily increasing in number, resulting in a radical change in the balance between young and old. This in turn results in new demands being made of urban mobility in future. Mobility is of central importance to this generation, which includes the benefits of autonomy, freedom and a high quality of life, thus enabling the maintenance of connections to family and friends, as well as enabling them to participate in wider society. This target group was described as being physically fit, active, enjoying travel, having large circles of friends and being open to new technologies. The group has an awareness of the need for social, economic and ecological sustainability. At the case study, four different urban scenarios have been presented, which should be investigated.

- Scenario 1: Shopping at the weekly market in the center of Berlin
- Scenario 2: Shopping at the mall
- Scenario 3: Trip into the countryside around Berlin e.g. a visit of allotment garden
- Scenario 4: "non"-barrier-free access to the railway station, using the example of Warschauer Straße

Six students from the architecture program and three research assistants participated in the two-day workshop. Afterwards, the participants had one week to work up the observations and present the results. The student workshop has been conducted in order to analyze the scenarios using the method cultural probes. "Cultural probes (also known as diary studies) provide a way of gathering information about people and their activities. Unlike direct observation (like usability testing or traditional field studies), the technique allows users to self-report" (Gaffney 2006). The main idea was to get in the given situation (see Fig. 6) and make observation with the help of: questionnaires, sketches, city maps, interviews and photos.

Fig. 6 Urban scenarios

After the field analysis, the gathered material has been discussed and evaluated during a workshop on the second day (see Fig. 7). In this workshop, many interesting ideas for PSS were discussed and developed. An example of this is the "pedal bus" (see Fig. 8). Similar to a regular bus, individual locations are approached and help and the acquired goods can be transported home. Moreover, the bus is powered by the passengers pedaling. On the one hand, this provides physical activity to the passengers, and on the other hand, an opportunity for social contact with others.

The main required customer values mainly considered for the further development of the use case were: social interaction, activity, mobility and communication. In a further expert workshop a transdisciplinary group of five researcher, including

Fig. 7 Workshop

Fig. 8 "Pedal bus" (Fillon and Klupsch 2013)

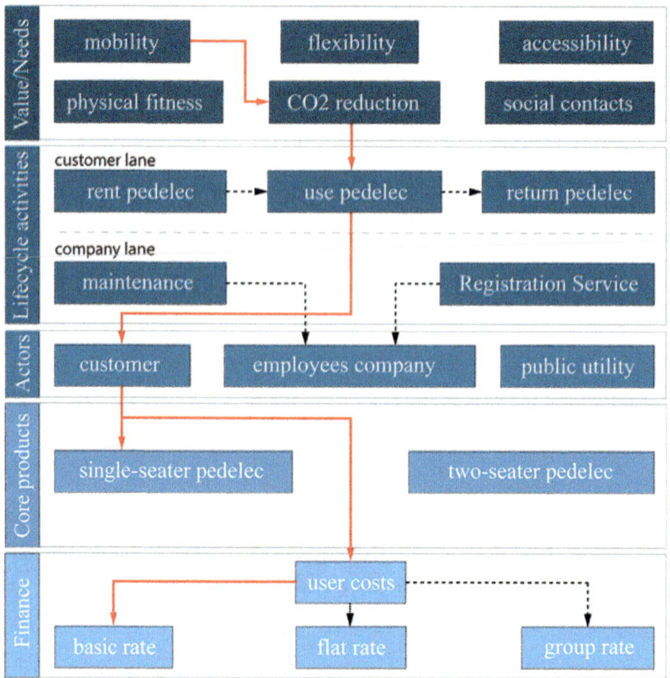

Fig. 9 Excerpt of a reduced use case with PSS layer method

one PSS expert, continued the idea of the "pedal bus" into different ideas for use cases. The core product is a pedelec (e-bike) with additional services and infrastructure elements. The PSS concepts have been developed using the PSS Layer Method (Müller 2014). An excerpt of a use case is provided in Fig. 9. The number of layers, the content and the linkage has been reduced in order to increase the clarity of the graphic. The following three scenarios can be described:

Scenario 1 describes the bike sharing concepts that already exists in similar formats. The user can rent and park the pedelec (one- or two-seater) at any station. A user platform, the Social-Pedelec-Network, is used to arrange joint meetings. At the stations, monitors are set up to ensure ease of use so that users can easily take advantage of the service. The customer has various options during the registration process, e.g. the selection of appropriate fares, which can be derived from the personal usage. A distinction is made between rates according to time, flat rate or group rates when multiple users want to perform a joint tour. A closing and registration system governs the billing and sending of opening codes. For information transmission and collection transmitters are used to derive from the information optimizations.

In Scenario 2, the customer has the possibility to obtain the pedelec at any location provided. A service staff brings the pedelec to the customer at the desired time and location. In this scenario, two-seater pedelecs are used exclusively. The service staff supports the customer in connection with various activities and returns with the customer to the desired location. The service staff takes care of both maintenance and servicing. The main objective is to support people with their daily need.

The third scenario particularly appeals to home communities. The pedelec stations are built near the residential communities. The station is at any time again terminated and degradable. Borrowing takes place, as in Scenario 1, a locking and registration system. Even in this scenario, the customer will have the opportunity to reserve pedelec at a customer platform with his user account and make loan requests. In this scenario, the pedelecs are additionally provided with a stair device. The stair devices are removable and facilitate users' horizontal and vertical mobility, e.g. the ability to climb stairs.

5.2 PSS Prototyping Methods

The development of the use case provides the research team with a profound basis of early PSS concepts developed with a well known and validated PSS design methodology. In a first step, a method is needed to extract test cases out from PSS concepts. Afterwards, a low and a high fidelity method are presented for concept validation.

5.2.1 Derivation of Test Cases from PSS Concepts

In order to ensure a systematic derivation of possible test cases of early PSS concepts as well as an integration of the validation dimensions and perspectives, a new method is required. For this reason a matrix has been proposed in Exner and Stark (2015). The matrix utilizes the common use of process flows used for the development of PSS, which is the input for the first column. The next two columns describe the related PSS elements to this process phase. The PSS Layer Method (Müller 2014) is preferred due to the already applied linkage between the elements. Nevertheless, the method is applicable to other PSS concept development methods. The prerequisite for the matrix is a concept that includes the process of the PSS as well as PSS elements. Eventually each process phase is assessed according to validation dimensions and perspectives (see Sect. 4). Following this step the development team can decide which phases have to be tested. Additionally, the matrix enables a systematic decision regarding the question: Which process phases and PSS elements can be tested with which PSS prototyping method? An excerpt of the matrix for the use case is shown Table 3.

Table 3 Derivation matrix for test cases (Exner and Stark 2015)

Process (customer view)	Services and software	Product, periphery and infrastructure	Validation dimensions	Validation perspectives
[...]				
2.4 Go to pedelec	Smartphone app, navigation (app)	GPS transmitter	Human-machine interaction, precision of navigation, [...]	Usability of the app with navigation (customer/developer)
2.5 Examine for damages	Checklist (app)	Pedelec, smartphone holder, GPS transmitter	Usability of the app, functionality and design of the pedelec, [...]	Usability of the checklist (customer/developer)
[...]				
2.9 Remove pedelec of charging station	Guidelines (app)	Pedelec, charging stations, smartphone holder	Usability of the app	Usability of the guidelines
[...]				
3.1 Defect while usage	Guidelines (app), provide help/alternatives (phone)	Smartphone holder, repair and transport infra-structure, customer service center	Usability of the app, driving properties, [...]	Support by unknown events (customer), process (developer)
[...]				

5.2.2 Development of UA4PSS

In a first step, validation methods of various disciplines have been analyzed (see Sect. 4) in order to identify existing validation methods for PSS, as well as assess common validation methods of other disciplines. A comprehensive evaluation method that integrates QFD, Consistency Matrix, Utility Analysis and VDI 2225 has been developed in order for the validation of PSS variants. Figure 10 illustrates the framework of this method.

A first qualitative testing indicated two problems regarding the usage of the method. Firstly, the method is too complex to be easily understandable. Secondly, the average time required for conducting the method has been measured to be almost three days. For this reason, a more user-friendly and expeditious method has been proposed (Exner et al. 2014c). The new method is based on the Utility Analysis and thus named UA4PSS. The method uses the first column of the derivation matrix. In a second step, evaluation criteria can be determined with the development team, e.g. with the brainstorming method. Finally, the process steps will be evaluated with the given criteria (effort, flexibility, complexity, safety and time) by a representative group of probands (see Table 4).

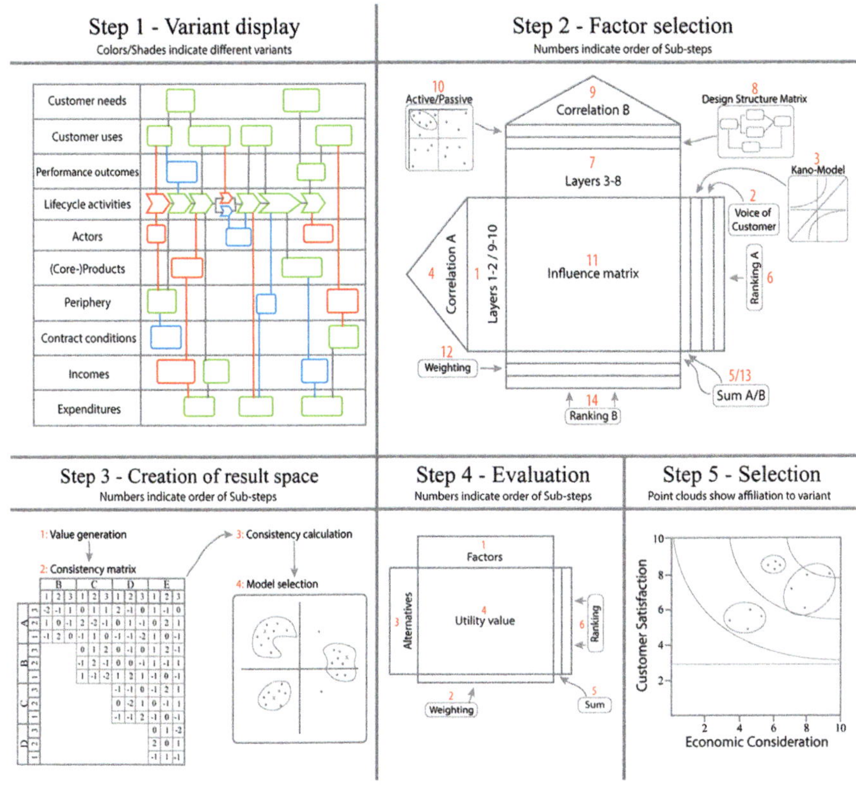

Fig. 10 Comprehensive variant analysis (Exner et al. 2014b)

Table 4 Evaluation matrix (Exner and Stark 2015)

Process phase	Criteria[a]		
	Effort	Safety	[...]
1. Booking of the pedelec	E.g. 10	E.g. −3	[...]
2. Open lock and remove pedelec of charging station			
3. Examination of damages			
4. Usage of the pedelec			
5. Malfunction while usage			

[a]Nomenclature
10 The criterion has a strong positive characteristic in this phase
6 The criterion has a positive characteristic in this phase
3 The criterion has a slight positive characteristic in this phase
0 The criterion has no effect characteristic in this phase
−3 The criterion has a slight negative characteristic in this phase
−6 The criterion has a negative characteristic in this phase
−10 The criterion has a strong negative characteristic in this phase
x The criterion cannot be assed with the given information

1. Booking the pedelec
You're already at the pedlec station. The booking process is conducted by entering the specific pedelec number in the smartphone app.
2. Open the lock and remove the pedelec from the charging station
The electronic locking system will open automatically by entering the specific pedlec number in the smartphone app. No further actions are necessary and the pedelec can be removed from the charging station.

3. Inspecting damages
Before use, you need to check the pedelec for any overt damages. The smartphone app offers a text field and allows for free text input for describing any potential damages. The information is subsequently reported directly to the service office.
4. Using the pedelec
Now you can mount your smartphone in the universal holder and begin to use it. While riding, you are constantly supported by an electric drive. Additionally, the smartphone app provides a navigation feature as well. You can interrupt your ride to take pictures of tourist attractions or other sights of interest.

5. Malfunction during operation
In the event of malfunction, the smartphone app features a text field and allows for free text input to describe the defect. The information provided is then reported directly to the service office. You will get an immediate reply to support you with the next steps to take, such as:
• To determine whether or not the pedelec can still be used
• To display the location of the next station
• To provide information regarding a potential refund due to the malfunction

Fig. 11 Description of the scenario for UA4PSS

In order to provide the necessary information for the probands, a summary and description of the process phases (see Fig. 11) is provided. The main idea is to repeat the procedure with several variants and the same criteria in order to evaluate the best solutions for each process step, thus enabling a selection of variants to be further developed.

The method has been tested with eleven PSS research experts in order to evaluate the feasibility of the method and the test matrix. The analyzed data indicate excellent results for 39 % of the ratings and only 6 % show a high statistical variance, thus are not statistically valid. The qualitative data and subsequent discussion reveal the reason in different understandings of the process phases and criteria for these ratings. Nevertheless, the overall usefulness of the method in order to assess early PSS variants has been proven (Exner et al. 2014c).

5.2.3 Development of SHP4PSS Concept

As described in Sect. 4, SHP typically consists of a physical prototype as an interaction device, digital models and a Virtual Reality environment. In order to cope with the most common PSS elements, e.g. services with procedural characteristics, periphery elements like smartphones and infrastructure, the SHP concept needs to be revised in order to develop a smart hybrid prototype for PSS (SHP4PSS). In reference to the use case and the extracted test case (see Tables 3 and 4) the required framework for SHP4PSS is represented in Fig. 12.

The main system elements of the PSS use case need to be represented with SHP4PSS in order to enable an integrated experience of product, service and infrastructure. Therefore, the concept focuses on the smartphone application, the pedelec, the parking station and the city. The smartphone application has been developed with the Blended Prototyping approach by Benjamin Bähr (cf. Bähr 2015). Furthermore, the integration of the smartphone app with the simulation has been planned in a two-step process. Firstly, the smartphone screen is streamed manually to the simulation computer and a response is initiated in the simulation by the operator. Secondly, an interface connects the smartphone application with the simulation and causes a corresponding effect in the simulation. The development of the physical prototype or hybrid simulator is being planned in a two-stage process as well.

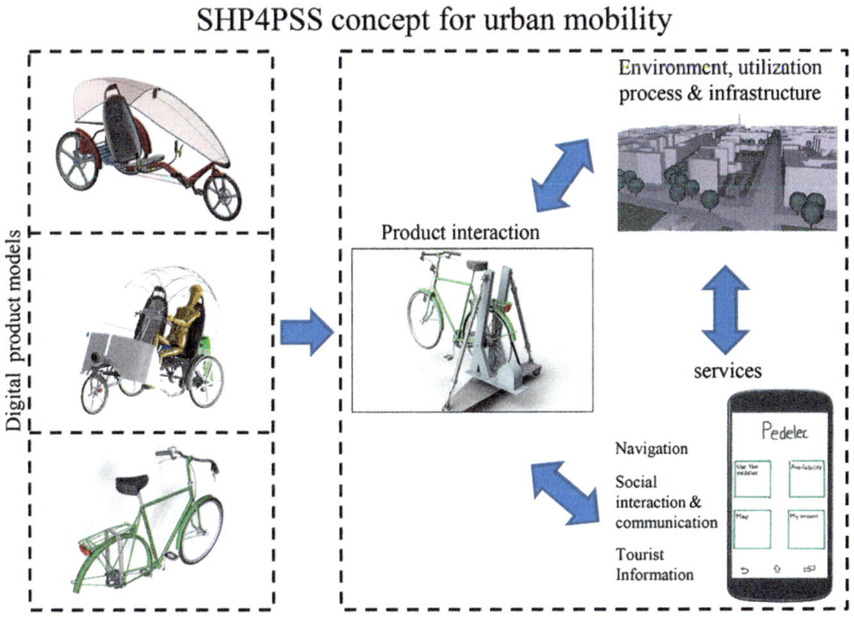

Fig. 12 SHP4PSS concept for an urban mobility PSS

Firstly, a device will be developed that integrates real product prototypes in a mechanism, including pneumatic muscles from FESTO, coupled with a magnetic particle break. This combination simulates necessary force feedback and enables a realistic experience. For this reason, basic user interactions like acceleration, steering and tilting will be measured with sensors. In a second step the device will be refined to a modular concept, thus omitting the need of an actual product prototype and enable a testing of different product variants with only the digital models needed. Regarding the environment, a digital city model of Berlin will be used to enable free movement in the scenario. Moreover, the model provides the opportunity to integrate infrastructure elements of the PSS, e.g. the parking station. Besides, digital product models will be integrated in the environment as well.

The critical aspect of the development and implementation of the components is the integration of all elements in order to ensure a realistic experiencing of the use case. Therefore, different software components have to be combined. The digital models will be implemented with the Unity 3D, the sensor data of the interaction device regarding the user input will be measured and transferred by an own software development and submitted to Unity 3D with an interface. The visual output of the simulation is converted with Techviz in order to provide stereoscopic projection in the CAVE.

5.3 SHP4PSS Development and Implementation

In order to provide the prerequisites for using SHP4PSS, several components need to be developed. The integration and implementation in VR requires the combination of specific software application. Afterwards, the SHP4PSS methods can be tested in a pre-study.

5.3.1 Development of the SHP4PSS Components

Smartphone Application: The smartphone application is in important part of the PSS as it enables the integration of different services, e.g. navigation, tourist information, communication in a user network etc. In a first step, the app supports registration, reports of damages and support while malfunction occurs. Nevertheless, additional functions have been prepared and can be integrated easily for further test cases. The development has been supported by the Blended Prototyping approach with a fast and easy sketching of the app framework. Afterwards, the created software code has been used to develop an optically improved version, see Fig. 13.

Digital Models: The development of digital models is a central aspect for the SHP4PSS approach. With respect to use case three, components need to be developed, such as a city model of Berlin, models of the product and the pedelec parking station. A rudimentary city model (level of detail 2) has been provided by

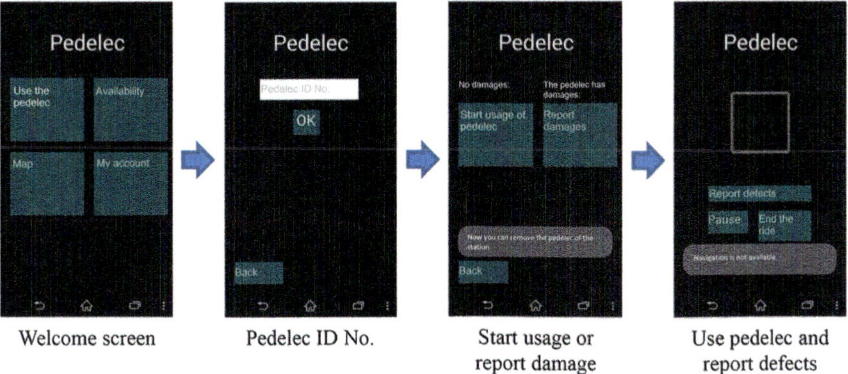

Fig. 13 Smartphone application

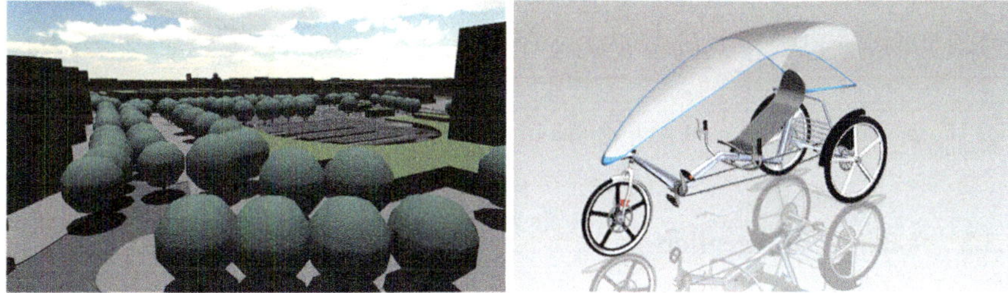

Fig. 14 Digital models

the city of Berlin. The textures, colors, shadows and sky have been optimized using Unity 3D in order to enhance the immersion in the later simulation. Furthermore, different models of the core product and the pedelec parking station have been developed. Some examples can be seen in Fig. 14.

Interaction Device: The interaction device has been planned with two pneumatic muscles, but was changed due to difficulties with controlling two active force inputs. The actual prototype enables tilting with one FESTO muscle and one spiral spring. The resistance while driving is implemented with a magnetic particle break and an electric engine, thus supporting a more realistic rolling resistance as well as driving up and down. Nevertheless, an actual tilting in this axis is not possible at this point. The data for speed, breaking, steering etc. is measured with several sensors and transferred to the computer program. In the first development stage of the interaction device a product prototype is needed which is attached to the rear part of the interaction part. The front part pivots the front wheel of the product prototype as well as serves as the parking station. Thanks to rear wheels, the product prototype can be taken off the station and parked again. Figure 15 shows an early development status.

Fig. 15 Interaction device used with oculus rift

5.3.2 Software Integration

The SHP4PSS demonstrator consists of several hardware and software modules. The modularity is in the sense of SHP. It allows easy and fast replacement of each module, and the CAVE, for example, can be replaced by an HMD like the Oculus Rift or with any other display technology. The architecture (see Fig. 16) not only describes the modules but also the data streams from and to each other. The user is the most important preset, but his role must be well defined because he closes the interaction loop between physical device and virtual environment. For this purpose he should be well integrated, ideally with his major senses (visually, auditory and haptic).

The general architecture can basically be separated into two different layers, the control layer and a visualization layer. While the control layer is mainly responsible for the processing of sensor input (steering angle, velocity, braking) and actor output (force feedback tilt) based on the physical interaction device, the visualization layer ensures an accurate, stereoscopic display of digital models and virtual environment. The most important technical interface in this architecture is the connection between control layer and visualization layer, where sensor commands coming from the interaction device and functional behavior of digital models need to be mapped accurately.

Six different modules, connected by seven data streams, represent the more detailed representation of the architecture. The physical interaction is realized by

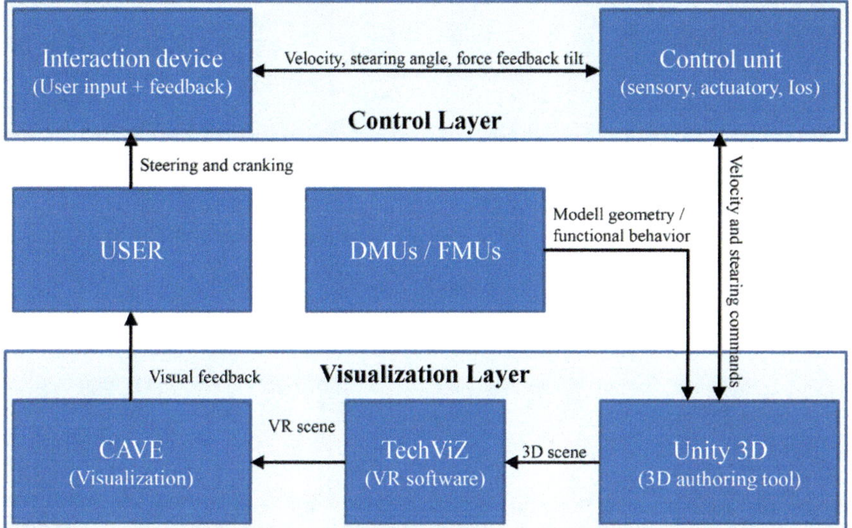

Fig. 16 Architecture of the SHP4PSS demonstrator

converting the user movement into control signals that need to be processed and translated into commands by the control unit afterwards. At this point the provided commands need to be connected to the applied digital models. Under these conditions, the digital models can be extended by an interactive behavior and need to be prepared for visualization. The visualization is mainly realized by using Unity3D. Here the functional models are integrated into a static, virtual environment. After extending the generated 3D scene by a stereoscopic view (TechViZ) the scene can be displayed inside a CAVE.

In summary, the user is now able to interact physically (enabled by control layer) with virtual models that hold functional behaviors within a mixed reality environment (visualization layer).

5.3.3 Pre-study with SHP4PSS

In order to receive quantitative results of the development status regarding the realistic experience of the SHP4PSS components, the project team used the event Hybrid Talks to present the work in progress. In 2014, the research team participated at Hybrid Talks and Showcase “Rethinking Prototyping” with a number of around 200 visitors presenting the driving simulation of Tacx, a digital city model with Oculus Rift as Virtual Reality, the concept of the interaction device and the prototype of the smartphone app (see Fig. 17). The researcher received mixed results due to the complexity of the approach, but the discussion of the separate elements indicated good feedback and useful comments for improvement.

Fig. 17 Hybrid talks 2014 demonstrating state of the art driving simulation

Fig. 18 Hybrid talks 2015—demonstrating the actual interaction device

In 2015, the research team once again participated in the Hybrid Talks "Human-Machine Interaction" with around 300 attendees. The main objective was to receive feedback regarding the interaction device. The prototype could be presented (see Fig. 18), but the team abstained from the use of Oculus Rift as VR due to common dizziness suing this technologies and thus probable injuries. During the showcase, approximately 30 visitors actually used the interaction device and reported a surprisingly good feedback, even without using the VR environment. The most intense discussion was focused on the correct tilting of the device in terms of the centrifugal forces. A consensus could be reached that there has to be a difference between the projection in VR and screen.

Eventually the complete SHP4PSS setup has been implemented in the CAVE (see Fig. 19). The experiencing and tilting could be adapted due to the tremendous feedback during the Hybrid Talks as well as an own pre-testing in the institute with several researchers and students. First qualitative results demonstrate a high immersion in the environment and a realistic experience of the driving process.

Fig. 19 SHP4PSS implementation in the CAVE

5.4 *Case Study*

The main objective of the case study is evaluating the feasibility of the SHP4PSS and UA4PSS methods. Moreover, the study shall provide reliable data in order to assess the cost-benefit ratio between the low fidelity and high fidelity approach.

5.4.1 Framework of the Study

The most important aspect of the research project is the evaluation of the feasibility of the new PSS-prototyping approach as well as a comparative evaluation with another validation method for PSS. A framework has been proposed in order to combine both perspectives (Exner and Stark 2015) (see Fig. 20).

In order to ensure quantitative results, a minimum of 12 probands for each method enable a statistical evaluation. Once the probands have completed the demographic questionnaire an explanation of the objectives and an exercise for increasing comprehension follows. For the low fidelity method the explanation of

Validation of test case with SHP4PSS	Validation of test case with UA4PSS
Test group: min. 12 probands each	
Demographic questionnaire	
Procedure: • Welcome and explanation of the objectives • Introduction to the Smartphone App with an exercise • Explanation of the test case • Experiencing of the test case in VR	Procedure: • Welcome and explanation of the objectives • Explanation and hand out of the visual and textual explanation of the test case
Evaluation of the test scenario due to given matrix and criteria Questionnaire regarding the application of the method (usability and acceptance)	
Empirical analysis and interpretation of the results	

Fig. 20 Framework of the evaluation for PSS-prototyping (based on Exner and Stark 2015)

Fig. 21 Case study of SHP4PSS and UA4PSS

the test case occurs on a purely textual and visual level, including pictures. For the high fidelity approach, a verbal explanation is followed by experiencing the test case in Virtual Reality. Subsequently, the probands evaluate the phases of the test scenario with given criteria and, both groups evaluate the method by filling out a questionnaire regarding usability, comprehensibility and hedonistic aspects. Finally, the results of both studies can be analyzed and compared. The hypothesis for the evaluation is: By using SHP4PSS the inter-reliability will be increased. Moreover, a main aspect of this analysis is the cost-benefit comparison due to the high variances of differences of effort in providing the test environment.

The evaluation was conducted according to the test case (see Table 4 and Fig. 11). Figure 21 documents the realization and completion of the case study.

5.4.2 Analysis of Collected Data and Results

The analysis of the data can be carried out in two different ways. Firstly, the assessment of the PSS concept of the probands according to Table 4 can be statistically evaluated. For this reason the mean value, variance and standard deviation (σ) determine the quality of the information provided by each method. It can be assumed that the lower the standard deviation, the better the information provision by the given method. This approach ensures an indirect analysis of the methods and minimizes the subjectivity of the probands. Secondly, the probands evaluated both methods regarding feasibility, usability and acceptance directly with standardized questionnaires.

Analysis and Interpretation of the Assessment of the PSS Concept: The assessment of the PSS scenario (see Table 4) consists of five questions regarding the five

Table 5 Analysis of the PSS concept assessment

Values	UA4PSS (%)	SHP4PSS (%
Very good results ($\sigma \leq 1$)	20	28
Good results ($1 \leq \sigma \leq 2$)	48	68
Poor results ($\sigma > 2$)	32	4
"Not rateable" (in proportion to all 300 ratings)	24	16

criteria for each of the five phases. The response options include a bipolar seven-stage scale interval and the possibility to choose "not rateable". Therefore, the interpretation has to be conducted while regarding two aspects. Firstly, it can be reasonably assumed that on a seven-stage scale a standard derivation of lower than one is very good. Furthermore, a value between one and two can still be rated as good. A standard deviation above two indicates too high a variance. This means that the clarity of the information provided is deficient. Secondly, "not rateable" means, that the probands have not received a sufficient input of information regarding the phase and criteria.

Table 5 indicates excellent results for SHP4PSS with 96 % good and very good results. Furthermore, only 16 % of all ratings have been "not rateable", whereas three fifth of these refer to one phase (usage). It can be argued, that only the driving has been implemented for the use phase so far. Therefore, with additional implementations (navigation, several driving modes etc.) the results would be even better. The results of UA4PSS are still good, but do not reach the level of SHP4PSS. The analysis reveals approximately one third of poor results and one quarter which are not rateable for UA4PSS. Furthermore, these results are evenly dispersed across the phases and criteria. These data display that although all necessary information has been provided, the probands had difficulties to imagine the concrete situations in the scenario. In that context SHP4PSS has a certain advantage by enabling a real experiencing of the PSS.

Analysis and Interpretation of the Feasibility, Usability and Acceptance: The questionnaire consists of a bipolar five-stage scale interval with eleven questions. The response options range from "not at all" over "neither" to absolutely. In order to enhance the clarity the results have been transferred to a percentage basis with absolutely as 100 %, see Table 6.

The results indicate an overall rating of good to very good for both methods. The differences regarding the usability are statistically irrelevant; each method has been rated with good. The feasibility and acceptance of SHP4PSS is with almost 90 %

Table 6 Results of the questionnaire (feasibility, usability and acceptance)

Category	UA4PSS (%)	SHP4PSS (%)
Feasibility (four questions)	Ø 73	Ø 88
Usability (four questions)	Ø 70	Ø 73
Acceptance (three questions)	Ø 62	Ø 89

quite high, thus the probands have been enabled to validate the test case and would use the methods for the development of PSS. Regarding UA4PSS the feasibility is good, but the probands are only slightly positive in accepting the methods as an important element for the PSS design process.

Finally, both parts of the analysis need to be considered conjointly. The overall results of the direct and indirect analysis indicate a positive evaluation of UA4PSS and SHP4PSS. Furthermore, SHP4PSS surpass UA4PSS in every category. The research team explains these results with the realistic experiencing of the PSS concept with SHP4PSS, thus the hypothesis mentioned before has been proven.

6 Conclusion

The main purpose of the "Hybrid Prototyping" subproject was to enable the prototyping of Product-Service Systems that could be seen as a desideratum at the outset of the project. The novelty of the concepts was the hybrid integration of digital models and physical prototypes and their utilization for PSS. The complexity of the intended concepts require a transdisciplinary approach due to the need for expertize drawn from various fields such as mechanical engineering, software engineering, user interaction, PSS development and urban areas, just to name a few of the more important ones. Additionally, due to the lack of a comparable method, the team also had to develop a low fidelity approach. The results could only be achieved through the assistance provided by other researchers within the departments, student projects, researchers from other subprojects and the excellent coordination of the "Hybrid Plattform".

It can be stated that the SHP4PSS approach accomplished the experiencing of early PSS concepts and enables the evaluation of different PSS variants with potential customers. Furthermore, PSS development teams can use this method in order to validate the PSS throughout the development process. Additionally, actual stages of the development process can be presented as milestones on a management level. Therefore, all three important perspectives (see Sect. 4) for the validation of PSS profit from this new method. Of course the cost-benefit ratio needs to be considered as well due to the high degree of effort in providing the SHP4PSS test environment. Clearly, a CAVE cannot be operated by small and medium sized enterprises (SME) nevertheless, when talking about PSS it is easy to think of providing an SME solution to develop PSS and conduct expert testing. For large companies already working with a VR, some could benefit from the knowledge and experience created. Furthermore, by developing a modular interaction device, the need for a physical prototype falls by the wayside, thus meaning that only digital models are really necessary to facilitate the test environment. The modular interaction device adapts to different forms of mobility solutions, e.g. a recumbent bicycle or a two-seater. A further idea that arose during the pre-study was to enable the device to tilt along the longitudinal axis, allowing for a better uphill and downhill experience. In summary, the initial investment in developing this solution seems quite high with respect to a simple use case. However, more use cases regarding urban mobility will benefit from the existing prototypes and infrastructure, and can thus be developed and implemented much more quickly and affordably in future.

In addition to the SHP4PSS method, the proposed UA4PSS approach is also worth considering. The low fidelity approach generated good results as well, but could not compete with SHP4PSS in terms of accuracy and a realistic experience. The quantitative and qualitative data indicates that a higher immersion and a realistic experience of the concepts create the necessary conditions for the validation of PSS concepts. The SHP4PSS user (the customer, developer or decider) attains a better comprehension of the concept while experiencing it in the later application. The user can thus empathize with the situation and the use case, thereby providing more detailed and accurate feedback. The project team proposes using UA4PSS for first PSS ideas in order to reduce the number of possible solutions before initiating concept development. In this way, UA4PSS will be utilized in the most efficient way and complete the methodology for early PSS development phases. Nevertheless, the project team plans to provide a third method between the low and high fidelity approach before the end of the project.

The project results have eventually to be compared with the research questions stated in Sect. 1.

- [RQ1] How can the demands of customers be analyzed to generate ideas and solutions for new mobility concepts in urban areas?

The PSS development process (see Fig. 5) initially focusses on the customer needs in order to generate ideas. Therefore, the customer has to be integrated in the PSS design process. Inspired by the Design Thinking mentality—to get in the real environment with the real customer—the method "cultural probes" has been applied. In this way, customer feedback and essential demands for new mobility solutions could be captured.

- [RQ2] How can PSS be prototypically implemented in order to validate PSS in early development stages, like planning and concept phase?
- [RQ3] How can the integration of the customer in the validation process be realized?
- [RQ4] How can the Smart Hybrid Prototyping approach be adapted in order to prototype PSS?

Regarding the prototypical implementation, two methods, a low and a high fidelity approach, have been developed and evaluated. Both methods enable a validation of PSS concepts in early development phases. Nevertheless, only the adaption and further development of Smart Hybrid Prototyping to SHP4PSS really integrates the customer in the design process. The user can interact with different PSS elements, thus increasing the reliability of the evaluation as well as providing the means for ad hoc design changes. To summarize, it can be stated that the project results answer all research questions satisfactorily.

Going forward, the transfer of knowledge to other cases or industries has to be considered. The research team reasons that SHP4PSS is adaptable to many cases working with mobility and urban areas due to experience gained in this project. The SHP4PSS infrastructure is adaptable and expandable and many building blocks already exist. As discussed in Sect. 4, a transfer of SHP regarding plant engineering and design is already the subject of current research. Due to the familiarity of the research team with these topics, in addition to mobility, further PSS cases could include services for machine tools and solar home systems. Further applications and the feasibility of SHP4PSS for the cases mentioned need to be discussed with researchers and industrial practitioners in workshops. Therefore, SHP4PSS has to be presented and discussed with industry developers and managers in order to review possible new cases, as well as acceptance in practice. The original SHP approach continuously received very good feedback from automotive industry practitioners and managers. Therefore, the research team is positive about the prospects for future workshops and options for the application in industrial research.

During the development of SHP4PSS, many impulses and feedback were received regarding architectural views in developing and presenting prototypes to customers. Of course, this research project is not intended as a one-way knowledge transfer. The possible application of the results for architecture can be determined in buildings with integrated services, e.g. shopping malls, in order to discuss procedures in the building with the client and potential customer of the building.

Finally, the research team would like to take the opportunity to thank all participants and supporters of this project. Without the significant assistance and contributions of these individuals, the project results would not have been so manifold and acknowledged.

References

Aurich, J. C., Fuchs, C., & DeVries, M. F. (2004). An approach to life cycle oriented technical service design. *CIRP Annals-Manufacturing Technology, 53*(1), 151–154.

Auricht, M., Beckmann-Dobrev, B., & Stark, R. (2012). Evaluation am Beispiel einer PKW-Heckklappe: Frühzeitige multimodale Absicherung virtueller Prototypen. *Zeitschrift für wirtschaftlichen Fabrikbetrieb*: *ZWF, 107(5)*, 327–331.

Bähr, B., Israel, J. H., & Exner, K. (2015). Perspectives on future prototyping—results from an expert discussion. *Rethink! Prototyping*.

Baum, M. (2008). *Urbane Orte. Ein Urbanitätskonzept und seine Anwendung zur Untersuchung transformierter Industrieareale*. Karlsruhe: Univ.-Verl. Karlsruhe.

Beckmann-Dobrev, B., Adenauer, J., & Stark, R. (2010) *Ein interdisziplinärer Ansatz zur multimodalen funktionalen Absicherung mechatronischer Systeme am Beispiel einer PKW-Heckklappe*. In Zusammenspiel von Maschinenbau, Elektronik und Software–Der Weg zum Gesamtfahrzeug: 3. Grazer Symposium Virtuelles Fahrzeug (pp. 18–27), Graz, Austria.

Blessing, L. T. M., & Chakrabarti, A. (2009). *DRM, a design research methodology*. London: Springer.

Bruhn, M. (2006). *Markteinführung von Dienstleistungen—Vom Prototyp zum marktfähigen Produkt*. In H.-J. Bullinger & A.-W. Scheer (Eds.) *Service engineering* (pp. 227–248). Berlin: Springer. doi:10.1007/3-540-29473-2_9

Burdea, G., & Coiffet, P. (2003). *Virtual reality technology*. Hoboken: J. Wiley-Interscience.

Burger, T., & Schultz, C. (2014). *Testen neuer Dienstleistungen: Ergebnisse einer empirischen Breitenerhebung bei Anbietern technischer Dienstleistungen*. Stuttgart: Fraunhofer Verlag.

Drewello, P. (2013). Embodied prototypes: The interaction of material system and environment. In C. Gengnagel, A. Kilian, J. Nembrini & F. Scheurer (Eds.), *Rethinking prototyping—proceeding of the design modelling symposium* (pp. 149–158). Berlin, Germany, Verlag der Universität der Künste, Berlin.

Engel, A. (2010). *Verification, validation and testing of engineered systems*. Hoboken: Wiley.

Exner, K., Buchholz, C., Sternitzke, A., & Stark, R. (2013). Validation framework for urban mobility product-service systems by smart hybrid prototyping. In C. Gengnagel, A. Kilian, J. Nembrini & F. Scheurer (Eds.), *Rethinking prototyping—proceeding of the design modelling symposium* (pp. 573–588). Berlin, Germany: Verlag der Universität der Künste, Berlin.

Exner, K., Hagedorn, L., Lindow, K., Stark, R., & Hayka, H. (2014a). Interplay of styling, design and service performance in developing product-service systems. In *International Conference on Engineering, Technology and Innovation (ICE), Bergamo, Italy* (pp. 408–416). doi:10.1109/ICE.2014.6871616

Exner, K., Lindow, K., Buchholz, C., & Stark, R. (2014b). Validation of product-service systems—A prototyping approach. In H. ElMaraghy, B. H. Windsor (Eds.), *Product services systems and value creation. Proceedings of the 6th CIRP conference on industrial product-service systems, Windsor, Ontario, Canada* (pp. 68–73).

Exner, K., Schnürmacher, C., Thiele, J., Schulz, F., & Stark, R. (2014c). Variants evaluation of product-service systems. In K. Brökel, J. Feldhusen, K.-H. Grote, F. Rieg & R. Stelzer (Eds.), 12. *Gemeinsames Kolloquium Konstruktionstechnik* (pp. 1–10). Bayreuth, Germany.

Exner, K., & Stark, R. (2015). Validation of product-service systems in virtual reality. In X. Boucher, D. Brissaud, & B. V. Elsevier (Eds.), *7th industrial product-service systems conference–PSS, industry transformation for sustainability and business* (pp. 96–101). France: Saint-Étienne. doi:10.1016/j.procir.2015.02.092

Fillon, B., & Klupsch, J. (2013). Workshop results: Pedal bus.

Gaffney, G. (2006). *Usability techniques series*. http://infodesign.com.au/. Accessed April 15, 2014.

Gengnagel, C. (2011). *Plusenergiehaus+E- Mobilität: Urbane Wohn- und Mobilitätskonzepte*. Heinze ArchitekTOUR2011: Architektur Bewegt. (Unpublished Lecture).

Hall, P., Pfeiffer, U., & Fischer-Schreiber, I. (2000). *Urban 21. Der Expertenbericht zur Zukunft der Städte. Menschen, Medien, Märkte*. Stuttgart: Dt. Verl.-anst.

Kossiakoff, A., Sweet, W. N., Seymour, S., & Biemer, S. M. (2011). *Systems engineering. Principles and practice*. Hoboken: Wiley.

Krause, F.-L., Anderl, R., Weber, C., & Rothenburg, U. et al. (2007a). CAD-DMU–FMU. In F.-L. Krause, H.-J. Franke & J. Gausemeier (Eds.), *Innovationspotenziale in der Produktentwicklung* (pp. 117–128). München: Hanser.

Krause, F.-L., Franke, H.-J., & Gausemeier, J. (Eds.) (2007b). *Innovationspotenziale in der Produktentwicklung*. München: Hanser.

Lindahl, M., Sundin, E., Shimomura, Y. and Sakao, T. (2006). An outline of an interactive design model for service engineering of functional sales offers. In D. Marjanovic (Ed.), *Proceedings of DESIGN 2006, the 9th International Design Conference, Dubrovnik, Croatia* (pp. 897–904).

Mannweiler, C. (2010). Einleitung. In J. C. Aurich & M. H. Clement (Eds.), *Produkt-Service Systeme. Gestaltung und Realisierung* (pp. 1–6). Berlin: Springer.

Martin, B., & Hanington, B. M. (2012). *Universal methods of design. 100 ways to research complex problems, develop innovative ideas, and design effective solutions*. Beverly, MA: Rockport Publishers.

Matzen, D. (2009). *A systematic approach to service oriented product development*. Ph.D. Thesis, Scandinavian Digital Printing A/S, Stokkemarke.

Meibodi, M. A., & Aghaiemeybodi, H. (2013). *Architectural "making" modes in relation to prototype notions*. In C. Gengnagel, A. Kilian, J. Nembrini & F. Scheurer (Eds.), Rethinking prototyping—Proceeding of the design modelling symposium (pp. 503–515). Berlin, Germany, Verlag der Universität der Künste, Berlin.

Meier, H., & Uhlmann, E. (2012). Hybride Leistungsbündel—ein neues Produktverständnis. In H. Meier & E. Uhlmann (Eds.), *Integrierte Industrielle Sach- und Dienstleistungen. Vermarktung, Entwicklung und Erbringung hybrider Leistungsbündel* (pp. 1–21). Berlin: Springer.

Milgram, P., Takemura, H., Utsumi, A., & Kishino, F. (1995). Augmented reality: A class of displays on the reality-virtuality continuum. In *Telemanipulator and Telepresence technologies. Proceedings SPIE* (Vol. 2351, pp. 282–292).

Miller, C. C. (2014). *Is owning overrated? The rental economy rises.* http://www.nytimes.com/2014/08/30/upshot/is-owning-overrated-the-rental-economy-rises.html?_r=0&abt=0002&abg=1. Accessed June 17, 2015.

Mont, O. K. (2002). Clarifying the concept of product-service system. *Journal of Cleaner Production 10(3)*, 237–245.

Müller, P. (2014). *Integrated engineering of products and services: Layer-based development methodology for product-service systems.* Berichte aus dem Produktionstechnischen Zentrum Berlin. Fraunhofer Verlag, Stuttgart.

Nguyen, H. N., Exner, K., Schnürmacher, C., & Stark, R. (2014). Operationalizing IPS2 development process: A method for realizing IPS2 developments based on process-based project planning. In H. ElMaraghy & B. H. Windsor (Eds.) *Product services systems and value creation. Proceedings of the 6th CIRP conference on industrial product-service systems, Windsor, Ontario, Canada* (pp. 217–222). doi:10.1016/j.procir.2014.01.024

Pahl, G., Beitz, W., Feldhusen, J., & Grote, K.-H. (2007). *Engineering design. A systematic approach.* London: Springer.

Pinto de Freitas, R. (2011) *Hybrid architecture: Object, landscape, infrastructure.* In: EFLA regional congress of landscape architecture. Mind the gap. Landscapes for a New Era (pp. 1–9). TAllin, Estonia.

Sadek, T. H. (2009). Ein modellorientierter Ansatz zur Konzeptentwicklung industrieller Produkt-Service-Systeme. Bochum, Ruhr Universität Dissertationen. Schriftenreihe/Institut für Konstruktionstechnik 2009, 1. Shaker, Aachen.

Sakao, T., & Lindahl, M. (Eds.). (2009). *Introduction to product/service-system design.* Berlin: Springer.

Sakao, T., & Shimomura, Y. (2007). Service engineering: A novel engineering discipline for producers to increase value combining service and product. *Journal of Cleaner Production 15(6)*, 590–604.

Schmid, M. (2005). *Service-Engineering. Innovationsmanagement für Industrie und Dienstleister.* Stuttgart: Kohlhammer.

Schneider, P., Clauß, C., Schneider, A., Stork, A., Bruder, T., & Farkas, T. (2009). Towards more insight with functional digital mockup 2009. In *Simulation for innovative design: Proceedings of the 4th EASC 2009 European automotive simulation conference* (pp. 325–336), Munich, Germany.

Shimomura, Y., & Arai, T. (2009). Service engineering—Methods and tools for effective PSS development. In T. Sakao & M. Lindahl (Eds.), *Introduction to product/service-system design* (pp. 113–135). Berlin: Springer.

Shimomura, Y., Hara, T., & Arai, T. (2009). A unified representation scheme for effective PSS development. *CIRP Annals-Manufacturing Technology, 58(1)*, 379–382.

Spur, G., & Krause, F.-L. (1997). *Das virtuelle Produkt: Management der CAD-Technik.* München: Hanser.

Stark, R., Beckmann-Dobrev, B., Schulze, E.-E., Adenauer, J., & Israel, J. H. (2009). *Smart hybrid prototyping zur multimodalen Erlebbarkeit virtueller Prototypen innerhalb der Produktentstehung.* In A. Lichtenstein (Ed.), Proceedins Der Mensch im Mittelpunkt technischer Systeme. 8. Berliner Werkstatt Mensch-Maschine-Systeme, (Berlin, Germany, 2009). Fortschritt-Berichte VDI: Reihe 22, Mensch-Maschine-Systeme 29. VDI-Verlag, Düsseldorf, pp. 437–443.

Stark, R., Israel, J.H., Wöhler, T. (2010). Towards hybrid modelling environments—Merging desktop-CAD and virtual reality technologies. *CIRP Annals-Manufacturing Technology, 59*, 179–182.

Stork, A., Wagner, M., Schneider, P., Hinnerichs, A., & Bruder, T. (2010) Functional DMU: Co-Simulation mechatronischer Systeme in einem DMU-Umfeld. *Fraunhofer Produktdaten-Journal, 1*, 44–48.

Suvarna, S., Stöckert, H., & Stark, R. (2010). *Project and design reviews in IPS2*. In T. Sakao, T. Larsson & E. Lindahl (Eds.), *Industrial product-service systems (IPS^2): Proceedings of the 2nd CIRP IPS^2 Conference* (pp. 307–3014). Linköping, Sweden: Linköping University.

Tan, A. (2010). *Service-oriented product development strategies*. Ph.D. Thesis, Scandinavian Digital Printing A/S, Stokkemarke.

Ulrich, K. T., & Eppinger, S. D. (2012). *Product design and development*. New York, NY: McGraw-Hill Irwin.

van Halen, C. (2005). *Methodology for product service system innovation: How to develop clean, clever and competitive strategies in companies*. Assen: Van Gorcum.

van Husen, C., & Meiren, T. (2008). *Mit systematischer Dienstleistungsentwicklung zu hochwertigen Angeboten*. In Technologie und Dienstleistung: Innovationen in Forschung, Wissenschaft und Unternehmen. Beiträge der 7. Dienstleistungstagung des BMBF, I. Gatermann, Ed. Campus-Verl, Frankfurt/M., 59–64.

Blended Prototyping

Benjamin Bähr and Sebastian Möller

Abstract This chapter summarizes the research carried out in the project "Blended Prototyping". It surveys prototyping mechanisms that enable early testing of user interfaces for apps, mobile applications running on smartphones or tablets (apps). A new prototyping approach has been designed, which provides groups of app designers with mechanisms to use paper sketches as a basis, for a quick creation of mobile app prototypes in group work. The approach primarily addresses early design stages, but includes processes to build more complex prototypes as well, which can be applied in later development phases. Tools were designed, developed, and tested that allow designers to use the approach in productive prototyping sessions. The development of the "Blended Prototyping" approach was shaped by feedback we gained from fellow designers, industry experts, scientists, and amateurs. The collaboration in the research project "Rethinking Prototyping" taught us new aspects and views on prototyping that found their implementation in the "Blended Prototyping" idea. This chapter summarizes this journey and explains the motivation and concept behind the approach. It demonstrates the use of the implemented tools and tells the story of their development. Results from two studies we conducted in the course of the project are shown. The lessons learned from these can help in the development of prototyping tool in the future.

Chapter Overview This chapter is structured into five parts. The explanation of the motivation for a new prototyping platform, and its relation to existing approaches, are provided in Sects. 1 and 2. The basic research and development approach we followed in the project is briefly addressed in Sect. 3. The project results are then displayed primarily in Sect. 4. Here, the development of the platform and its usage are explained. Two different studies to evaluate "Blended Prototyping" are displayed

B. Bähr (✉) · S. Möller
Quality and Usability Lab, Technische Universität Berlin, Berlin, Germany
e-mail: baehr@cs.tu-berlin.de

S. Möller
e-mail: sebastian.moeller@telekom.de

C. Gengnagel et al. (eds.), *Rethink! Prototyping*,
DOI 10.1007/978-3-319-24439-6_9

and their results discussed at the end of the Sect. 4. A discussion of the research project results in general, and of further work specifically, is presented in Sect. 5.

Remarks on Terminologies "Blended Prototyping" is a design tool that aims to bring together people of different professions to explore interface ideas collaboratively. Therefore, the roles of designers and developers can often not be strictly distinguished in this text. It is the purpose of the platform to provide an environment that teams can use regardless of their professional background. Furthermore, in the discussion of development tools, two different kinds of references have to be made to the role of a user. First, there is the user of the development tool, who is a designer or developer. Second, there is the user of the prototype, who serves to provide feedback on a design idea. Please bear these terminologies in mind when reading this text.

1 Motivation of User Interface Prototyping

The term prototype is an aggregation of two ancient Greek words: *proto*—meaning first, and *typos*—meaning gestalt, or shape. Therefore, it can be understood as a preview version of a product that is not yet entirely developed. Prototypes are needed to put an idea into a form where it can be viewed and tested and thus generate insights for the further development. As displayed in the "Rethinking Prototyping" research project, prototypes are widely applied, in different fields and in different forms.

Prototypes are widely used, particularly in software development. The Institution of Electrical and Electronics Engineers (IEEE) takes the perspective of computer science in describing the process of prototyping as "a type of development in which emphasis is placed on developing prototypes early in the development process, to permit early feedback and analyses in support of the development process" (Radatz 1997). This is particularly true for the user interfaces of software, which cover a big part of the interaction space between user and machine. It is today's conventional wisdom, that successful software development is characterized by repeated tests and iterative refinement (Nielsen 1993; Szekely 1994). This is good advice, especially in early design stages. Here, principle design decisions have to be made that have essential impact on the success of whole development project (Winters et al. 2004).

However, in early design stages the product typically exists in the form of ideas and concepts, which are not well presentable to the user. Standard development tools, like programming environments, are too complex and thorough to come up with quick testable results. Therefore, approaches are needed to support a fast iterative prototype driven development process.

The idea of prototyping is not just valid for the development of standard computer software, but particularly for user interfaces of mobile applications. User interfaces for mobile apps have to deal with a number of additional challenges that derive from their changing use contexts. Unlike stationary software, mobile apps are not exclusively used sitting at a desk; but in the subway, walking on the street, dancing in a club, and in a myriad of different other contexts. Factors that are capable to affect the interaction strongly in different usage scenarios are for instance lighting conditions, surrounding noise, or the user's movement and posture (de Sá and Carriço 2008a, b). To handle these challenges appropriately, iterative tests of mobile applications with prototypes have to be conducted.

2 Applied Concepts and Related Work

This section explains the concept development of "Blended Prototyping". It briefly points out the range of existing prototyping styles and explains the position of "Blended Prototyping" in this context. The approach adapts ideas and advantages of paper-based prototyping, which is consequently displayed in more detail. Followed by that, in Sect. 2.2, "Blended Prototyping" is compared to other approaches that facilitate user interface prototyping of mobile apps.

2.1 Discussion of the Concept Behind "Blended Prototyping"

2.1.1 Evolutionary Versus Throwaway Approaches

Two opposing prototyping paradigms can be put in contrast, which deal in different ways with the utilization of the prototype after the test (Szekely 1994). *Throwaway* prototyping approaches see the prototype's value solely in the testing. After a prototype was tested, and the lessons-learned are noted, the prototype itself looses its value and can be thrown away. The prototype of the next iteration cycle will only be based on insights gained. In contrast, *evolutionary* prototyping approaches reuse the prototype itself after the testing. Here, the new prototype is created from an altered version of the old one. Throwaway approaches are typically used in early design stages, where the system's requirements are not yet clear and evolutionary approaches would bear too much risk to waste efforts in the wrong direction. However, after a sufficient number of iterative tests, the requirements become precise enough to move to evolutionary approaches.

Evolutionary prototypes are usually built with the same tools as the later software product. In contrast to that, throwaway prototyping approaches use special tools and techniques to help designers to quickly come up with a testable prototype. *Paper Based Prototyping* is an example for a throwaway prototyping technique, which simplifies the production of testable prototypes to a radical extent.

2.1.2 Paper Based Prototyping

The Paper Based Prototyping approach (PBP) is described in detail by (Snyder 2003). Here, a short highlight of the principle idea behind the concept is provided, and its most important advantages are pointed out. Later, in Sect. 2.1.4 we describe how "Blended Prototyping" adapts the approach and facilitates its advantages for the collaborative design process and prototype testing.

The core idea of PBP is to enable quick and easy design of prototypes by changing the prototyping media from computers to physical paper. It uses paper in the design process, as well as in the test of the prototype.

In a paper based prototype test, a paper version of a user interface is presented to a test user. The user is now asked to interact with the paper as if it was a real piece of software. A design team member, playing the role of the computer, will change the paper interface in reaction to the user's input, much in the way that the real software would reply. How this is done, is left open to the phantasy and practices of the design team. For example, small paper keyboards can be introduced, gestures can be regarded, and even animations can be simulated. The user typically provides comments on his activity in a think-aloud protocol, where he speaks out his thoughts and plans for everyone to hear. A test session is usually supervised by at least three team members, one playing the role of the computer, one writing a protocol with comments on his observations and a third one, acting as a moderator towards the test user. Ideally, paper based prototype tests are situated in usability labs. Here, the tests can be recorded with video cameras, allowing for a later analysis of the experimental data.

PBP—Advantages for the Design Session

The prototype is designed on the basis of paper with regular physical design tools like pens and scissors. No expert knowledge is needed to participate in a paper prototyping design session. Therefore, team members with different professional background can participate equally in the design sessions. Using pen and paper is generally perceived to be a natural tool to express ideas without obstacles. Computer oriented design tools risk hindering the free flow of creative thinking; they limit the degrees of expression to the functions implemented in the software (Bailey et al. 2008; Cook and Bailey 2005; Klemmer et al. 2001). Paper is patient.

Moreover, discussing and designing on the basis of paper has positive effects on teamwork (Beyer and Holtzblatt 1997; Klemmer et al. 2008). Aspects that come naturally in collaborative sessions with physical objects are hard to achieve in computing systems. Examples for this include dealing with personal and public space, sharing of content, or simultaneous manipulation of objects in a group.

Maybe the biggest advantage of the PBP approach is its speed. Snyder estimates, that in the prototype development with standard computer tools, the time share of programming is about 90 % (Snyder 2003). In PBP sessions there is no programming, however, additional effort exists in training a team member for using the computer later in the tests. In the end, especially for early prototypes, the time effort for PBP is usually significantly lower.

PBP—Advantages for the Prototype Testing

Paper based testing sessions have advantages towards those conducted with computer systems. Facing a handmade, rough looking paper version of software allows users to understand that they are not dealing with a finished product, but a pre-version they are asked to criticize. Users therefore tend to articulate their feedback about the prototype more freely (Snyder 2003). Moreover, the test user is personally accustomed to the tool used to create paper prototypes. This allows the design team to permit a test user make suggestions for alternative designs.

Schumann et al. (1996) observed that many architects present their early design ideas to customers with sketches rather than with photo-realistic rendered computer models.

In interviews they found that this could be explained by the ability of sketches to communicate certain design aspects in a more targeted way. A sketch highlights aspects by leaving out information.

This advantage can be utilized in sketched user interface prototypes as well (Klemmer et al. 2001; de Sá and Carriço 2006). If, for example, software is sketched in black and white, a test user usually gets the idea that colors are not relevant to the design team at the given stage.

2.1.3 Paper Based Prototyping in the Mobile Context

As mentioned above, PBP sessions should best be conducted in usability labs, where the test can be optimally monitored and recorded. Of course, mobile apps can be tested in these surroundings as well. However, the question has to be asked, whether a test in a stationary environment can give information on the specific challenges mobile interfaces have to solve. Mobile devices are used in an endless number of use contexts, which often bring their particular obstacles into the interaction (de Sá et al. 2008).

Kjeldskov et al. (2004) doubt the necessity of tests in the mobile context. In their work, which is entitled "Is it worth the hassle? […]", they point out the advantages of laboratory tests for optimally observing the user and emphasize the great effort mobile tests require. A direct reply to this point of view is provided by the title of a paper by Nielsen et al. (2006): "It's worth the hassle!". In opposition to Kjeldskov et al., many authors (Brewster 2002; Consolvo and Walker 2003; Duh et al. 2006; Monrad Nielsen et al. 2006) underline the need of mobile tests to find usability problems that derive from the mobile context.

Mobile user tests based on physical paper were conducted by de Sá and Carriço (2008a, b), who equipped users with wooden device dummies where cardboard screens could be inserted. However, in this test setup, the user had to be followed by design team members who had to catch up to play the role of the computer or observe the situation. This technique was not able to produce a natural use context, and a number of specific mobile interaction challenges could not be investigated: Interacting with cardboard cannot simulate low touch precision, nor can issues like reflection on the screen be considered. Therefore, many authors, including the ones of the mobile device dummy study, express the need to conduct mobile tests on the same mobile device, the finished product will be used on (Kieffer and Vanderdonckt 2007; Kieffer et al. 2010; de Sá et al. 2008).

2.1.4 "Blended Prototyping" Design Paradigms

When we thought about the concept of "Blended Prototyping", our primary goal was to address the needs of mobile development teams. Hence, we talked to different professionals from the Berlin startup scene about their specific development conditions. The picture emerged that apps can only be successful when they manage to provide a smart and creative solution to a specific problem, at a specific time. Therefore, apps are usually developed in small but effective teams, in companies that are oftentimes rather small that have to deal with different resource shortages. We learned that initial app ideas were explored and discussed in creative session that involved the whole team. And though there are apps that deliver a tremendous financial outcome, most projects do not become success stories.

Therefore, we wanted "Blended Prototyping" to become a tool that is designed to support collaborative creative thinking. It should produce prototypes as quickly and easily as possible, and thus put design teams in a position where they could establish a design process that is based on iterative tests. As displayed above, Paper Based Prototyping is a suitable approach to address these objectives. "Blended Prototyping" thus aims at adapting the advantage of PBP for the design process and test of user interface prototypes. At the same time, however, the approach should implement mechanisms that make the prototyping results testable directly on mobile devices.

As discussed in Bähr and Neumann (2013), "Blended Prototyping" uses physical paper, to best utilize the advantages of Paper Based Prototyping for the design process. Different authors compare the use of paper to digital substitutions, like stylus tablets. The found that physical paper allows a faster and more natural interaction in groups, and benefits collaborative brainstorming processes (Bailey et al. 2008; Beyer and Holtzblatt 1997; Newman and Landay 2000). To employ paper in a computing system, we decided to use an overhead projected tabletop system that can combine physical paper content with digital projections. Tabletop computing environments are frequently used to build collaborative systems with tangible objects, e.g. (Spindler et al. 2009; Underkoffler and Ishii 1999; Zufferey et al. 2009).

"Blended Prototyping" is designed to be as unobtrusively as possible so as not to interfere with creative teamwork. It can be used for developing simple prototypes on the basis of paper sketches as fast as possible. At the same time, however, designers can add program code to a prototype, thereby advancing the functionality of a prototype as complex as needed.

2.2 *"Blended Prototyping" in the Context of Other Prototyping Approaches*

The idea to use sketches in computer tools for user interface design is not new. The SILK framework was introduced (Landay 1996) already in 1996, a software enabling user interface design on the basis of sketches that are created using a mouse with a desktop computer. On the sketches, areas were defined that invoked the change of the interface on another screen. The finished design was then tested as a click dummy in a mockup player. SILK was developed further into DENIM (Newman et al. 2003), which focused the design on electronic tablets and introduced semantic zooms.

With the advent of mobile devices, new platforms for the sketch based design of user interfaces were explored. de Sà and Carriço (2008) developed an approach that used mobile devices and styluses for the sketch design. To allow autonomous user test in the mobile context, they implemented first logging mechanisms, which recorded the user input in the test.

A whole new world of tool-supported collaboration is explored with tabletop computing systems and interactive surfaces. From the start on, these approaches embedded physical objects like paper into the interaction scope. The DigitalDesk (Wellner 1993) created a work desk scenario, where office tasks were done on regular paper sheets. Other approaches investigated the collaborative interactions in working with 3D objects (Piper et al. 2002; Underkoffler and Ishii 1999; Zufferey et al. 2009) or transferred the space above the tabletop into a three dimensional interaction space (Spindler et al. 2009). With more affordable tabletop computing systems, more and more applications were developed that facilitate collaboration, e.g. (Battocchi et al. 2009; Ringel Morris et al. 2006; Shen et al. 2002; Tuddenham et al. 2009). Klemmer et al. created a system that uses an interactive wall with paper sticky notes as an environment to do collaborative prototyping (Klemmer et al. 2001). However, the application is not targeted at user interface design, but at a collaborative information structure prototyping of websites.

Commercial tools that allow early prototype tests on mobile devices exist. Well advanced tools for purchase can be found in Axure (2014), MockFlow (2014), or Balsamiq (2014). A free alternative can be found in the appInventor, which resulted from cooperation between the MIT and Google. These tools follow the fashion of traditional interface builders: Designers use computer software to create interfaces by positioning and connecting standard controls on a design stage, without much need to do programming.

Other commercial approaches leverage teamwork by avoiding PCs, but putting paper and mobile devices in the center of the design process. Apps like POP or the MarvelApp let designers first draw their interface ideas on paper, which are then photographed with the mobile device camera. Now, on the mobile device, areas are defined that link the single interface sketches to another. Holzman et al. (2012) follow a similar approach, but substitute the photographing and editing on the mobile devices by using the Anoto (2014) technology in the paper design process.

By tracking the position of Anoto pens in the drawing process, the results are automatically transformed into the digital space. Mobidev (Seifert et al. 2011) uses pre-trained shapes in the drawings to automatically recognize controls. Moreover, the system allows designers to enrich the prototype's functionality by entering code snippets directly on the mobile device.

Like other tabletop computing tools, as well as the commercial prototyping apps described above, "Blended Prototyping" focuses its design process on physical paper. However, as opposed to these approaches, "Blended Prototyping" allows a developer to implement complex interface behavior. Most prototyping tools discussed above produce results that are limited in their functionality to mere click-dummies. The mentioned commercial interface building tools offer visual processes to implement basic prototype behavior. Some other approaches support programming of functionality with code (Dalmasso et al. 2013; Smutny 2012), however, they all use web-technologies and pseudo-code. The use of web-technologies offers the advantage that prototypes can be tested easily on different platforms. Compared to the native programming process facilitated in "Blended Prototyping", such techniques are however limited in their capabilities and rights. Moreover, code produced with such tools cannot be integrated in the later programming, but has to be completely rewritten.

A number of approaches address this problem by providing remote mobile Wizard-of-Oz prototype test sessions (Davis et al. 2007; Klemmer et al. 2000; Segura and Barbosa 2013). Here, a test user tests the prototype in the mobile context, however, a member of the design team manipulates the app behavior live and remotely, acting as a kind of Wizard-of-Oz from Baum's novel (Baum and Denslow 2014). This allows for an easy simulation of even complex behavior. However, the degree of effort involved in the testing is very high.

3 Research Approach

The key questions we address in our research are the following: How can designers and developers be supported, to evaluate their design ideas on mobile user interfaces in early development tests? Which specific needs should be met in this context, to motivate a design process, based on prototype driven tests as early as possible? How can such requirements be translated into design paradigms for prototyping tools? And finally, to provide a basis to compare the performance of different tools: How can we measure the success of prototyping tools to meet these requirements?

The purpose of prototyping is to quickly put ideas into a more concrete form, where they are better accessible to others. This way, they can be tried and discussed at a stage early enough, to regard changes without time extensive principle revisions.

In our research, we adapt this basic thought of prototyping. Therefore, from early conceptual phases onwards, we repeatedly discussed and adapted our approach. We pursued the scientific discourse, we talked to mobile interface designers and developers, and we tested the clarity of our concept in demonstrations to the general public.

We quickly realized that useful prototyping approaches should support and encourage useful design processes. "Blended Prototyping" is targeted towards the user centered "Usability Engineering Lifecycle" by Jakob Nielsen (1993). It demands an iterative development process, which produces prototypes in each iteration cycle. The prototypes are then tested, which lead to results that are regarded in the next iteration. Nielsen underlines the relevance of a competitive analysis, where different user interface ideas are put to life in a prototype and then tested in comparison. The iteration cycles should be kept as short as possible, preserving the chance for inexpensive corrections. The Usability Engineering Lifecycle served as an orientation for providing tools in our approach, which address the single prototyping phases: the requirements analysis, prototype design, testing, and the interpretation of the test.

The research on "Blended Prototyping" considered topics from different scientific fields. These included questions on computer science, human computer interaction, design research and the investigation of collaborative creative work. Only an interdisciplinary team of researchers is able to tackle such a broad research field.

In our research we identified areas of specific personal interest, which we worked on independently, but never in an isolated manner. We conjointly discussed and united our results in short iteration cycles.

Within our work, we tried to achieve an understanding of how people of different professional backgrounds address issues in different ways. We wanted to open up personal toolboxes and learn to adapt and include the mechanisms applied by our partners. This exchange not only regarded the interdisciplinary work within the "Blended Prototyping" research project, but also the comprehensive "Rethinking Prototyping" research project as a whole.

4 Development and Evaluation of Blend Prototyping

4.1 Feedback Driven Development

As explained in the previous section, recurring feedback loops shaped the development of the "Blended Prototyping" platform. This generated insights, encouragements and critiques, which proved very valuable to our design process. Due to space limitations, this text cannot go into greater depth to review all the lessons we learned this way. It is rather meant to provide an impression about the diversity of both the experts and the amateurs we addressed.

In the early conceptual stages of the system, feedback was primarily gained from discussions with colleagues. This included workmates from the Quality and Usability Lab (Technische Universität Berlin/TU Berlin) as well as the Design

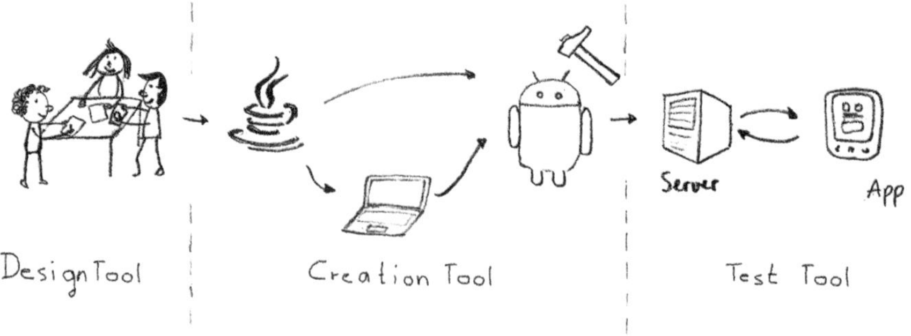

Fig. 1 Overview—"Blended prototyping" platform

Research Lab (Berlin University of the Arts/UdK Berlin). Here, principle questions about the approach, as well as technical details, were at the heart of discussion.

Thus, early demonstrations of the system with amateur youths were held in the years 2012 and 2013. For this we visited a programming workshop on Android for 10th grade school children in Freiberg, Saxony. Here, we tested how young novice users were able to understand and use the "Blended Prototyping" environment. Other forms of demonstrations and try-out sessions with amateur users were done in the context of the "Lange Nacht der Wissenschaften" (Long Night of the Sciences) in the years 2012 and 2013. At these occasions, a "Blended Prototyping" design tool was exhibited for a whole night to a large number of guests of different ages and professional backgrounds.

Additionally, we discussed the approach and its technical implication at different scientific conferences. Here, we addressed the communities of the MobileHCI (Int. Conf. on Mobile Human Computer Interaction), the CHI (Int. Conf. on Human Factors in Computing Systems), and the DSMB (Design Modeling Symposium Berlin).

"Blended Prototyping" was also subject of two large studies that we conducted in 2014 and 2015. In Sect. 4.3, we present results from discussions with 15 experts whom we invited individually to discuss tool requirements and the implementation of the platform. In the following Sect. 4.4, we highlight results from a user study wherein we surveyed the productive collaborative use of three different prototyping tools for working on design tasks (Fig. 1).

4.2 System Usage

The "Blended Prototyping" platform is structured into three modules that support all phases of the prototyping process, from the initial design to the prototype test. This section gives an overview of the different supported prototyping processes.

References to more detailed descriptions of single aspects that were published previously are provided in the corresponding segments.

The first module of "Blended Prototyping" supports the design process with a tabletop computing setup where groups can meet, discuss and shape mobile user interface ideas on the basis of paper sketches. The second module provides processes, for converting the data generated in the first step into a software prototype, which can be run on Android devices. The third module offers an infrastructure for conducting tests with the prototype, even with a bigger numbers of users. It includes tools for the management of test subjects, as well as for the analysis of the data, generated in user tests. The data shared between the modules is handled in files, which follow an open standard JSON logic.

Module 1—Design Tool

The "Blended Prototyping" design tool is set up around a regular table. Above the table surface, a video projector and a photo/video camera are installed which serve as input and output channels to a computer that controls the setup. A more detailed description of the setup is given in (Bähr and Neumann 2013) (Fig. 2).

Designers meet at the table to discuss and progress their user interface ideas. The interface is designed on regular physical sheets of paper. Designers can draw their ideas, build collages, or apply whatever techniques they like. The paper sheets have an imprinted barcode marker. This marker is recognized from the video signal, so that the system can determine the exact position of each sheet of paper lying on the table.

The "Blended Prototyping" system aims to be as unobtrusive to the design process as possible. This means that its users are free to decide when they want to use certain system functions, and when they prefer to ignore the system and concentrate on their design discussions and drawings. At the start of a design session, the system is typically ignored to a large extent. Later, when the interface prototype gets more complex, the design team will usually use the computer system more frequently.

Fig. 2 System overview

Interaction Within the Tabletop Environment

The mechanisms for the users' interaction with the "Blended Prototyping" platform were revised multiple times in the research project. As a first draft, standard input devices were used: a wireless computer mouse and a keyboard. However, this input channel undermines the collaborative system use, since only one user at a time has the power to control the environment. Moreover, in the context of a tabletop environment that is used from different sides of the table, the use of a mouse faces orientation problems.

Hence, in the following version, mouse and keyboard were excluded and mobile devices came into play. Since then, multiple tablet devices are commonly used that allow multiple designers to access the system at the same time. Wirelessly connected to one another and with the design tool, the data on the prototype is shared with an encrypted JSON object based communication via web-sockets. This solution works in a stable way and supplies a well-advanced input precision of the mobile devices. Moreover, the mobile devices can be used to view and test the prototype instantly on the mobile device.

However, compared to low fidelity tools, mobile devices can have negative effects on collaborative work, mainly in reducing the awareness of the group members to one another (McAdam and Brewster 2011; Rädle et al. 2014). During the time of usage, the device user will not actively participate in the discussion, nor will his teammates have the chance to participate in what he is doing. We accordingly addressed this issue and minimized the time needed for the single interaction tasks.

To avoid the use of mobile devices in the tabletop interaction, we developed a number of low-fidelity alternatives. Here we used colors and patterns that are recognizable by image algorithms. That way, controls could be added to the sketches, simply by drawing the according marks. Furthermore, we used colored rubber cords to define interface transition paths. However, in the current state these approaches are not reliable enough to make mobile devices in the interaction processes redundant.

Process of Digitalization/Mixing the Physical and Digital Space

Just like in the Paper Based Prototyping process, a prototype is created out of the sum of different interface states or screens, each of which is drawn on a single sheet of paper. Whenever the designers feel confident about a sketch, they can use the system to transform the physical into a digital version. For this, the camera of the setup takes a digital photo of the table surface, from which the sub image of the paper sheet is separated. After being photographed, the paper sheet is substituted with a blank version with the same marker. Now the picture of the former physical content is displayed on the paper as a projection. Users can draw additional content into a projection. In the next digitalization step, this new physical content will then be added into the digital interface version. Each interface photo at each decision step is stored separately in the computer system. A user can browse through the images of the single interface elements and is free to rollback the interface sketch to a previous version. This way, new features can be quickly added and tried out. However, they can also easily be discarded, without having to draw the whole screen again from scratch. Single interface screens oftentimes share a lot of content, like page menus or tab structures. The system provides mechanisms to copy such content from one screen to another.

The platform uses paper sheets of two different sizes: DIN A4 pages, which display screens in 8″, and half-sized DIN A5 pages that display a 4″ screen size. Different sized papers with the same screen number sync their digital content. Whether a designer prefers to use the smaller or bigger pages is left up to him. Bigger pages might provide a better reference to draw tablet apps, where smaller pages are better suited for apps that are designed for smartphone screens. However, designers might want to choose the 8″ version to design a smartphone app, since it offers more space to create complex drawings. On the other hand, more small pages fit onto the table at the same time, which helps provide a better overview of the designed interface.

Defining User Controls

The "Blended Prototyping" design tool supports different user controls, which build the foundation for the later prototype functioning. Currently, the system supports two custom controls, buttons and gesture listeners, as well as a number of standard Android controls: radio buttons, check boxes, text boxes, image containers, and video players. Gesture listeners are controls that are able to recognize touch gestures, to provoke an interface change. In this way, the prototype can recognize a swipe gesture to trigger an interface screen change, for example.

Designers can position controls within the interface sketch where they will be rendered in as an element in the prototype that the user can interact with. Buttons and gesture listeners are invisible to the prototype user; they are supposed to be identified by the user from the interface's sketch. The other controls are displayed in the same high-fidelity design as the according standard Android controls.

Depending on the nature of the user controls, supplemental information regarding the control position and size can be defined in the design tool. For a button or gesture listener, the interface screens it may link to are defined. The gesture listener can be set up to listen for different kind of trigger gestures. For radio buttons, the grouping is an important additional aspect.

Within the tabletop design tool, user controls are highlighted in the digital interface projections. Interface screen changes that are invoked by a control are displayed as animated lines that connect the control to the target screen. This way, the interface's transition paths are highlighted the tabletop surface in the form of a storyboard view, as illustrated in the middle picture in Fig. 3.

Module 2—Creation Tool

The following text provides a brief overview of the mechanisms to process data from the design tool to a prototype that can be run on a mobile device. A more in-depth description of the matter can be found in Bähr (2013).

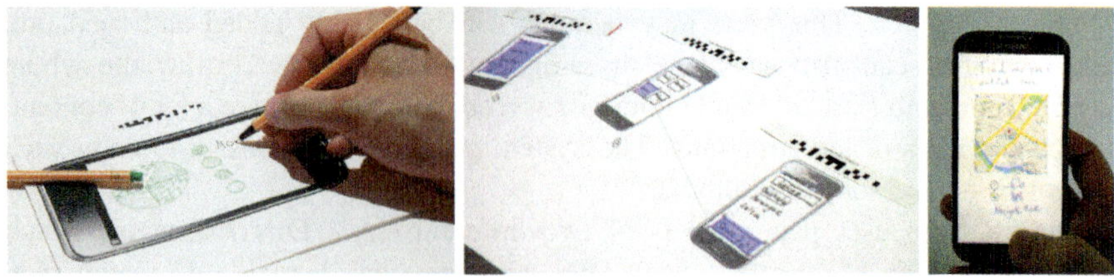

Fig. 3 Blending physical and digital content (*left*); storyboard view of an interface in the design tool (*middle*); Running prototype on mobile device (*right*)

The platform offers two processes to generate prototypes from the data generated in the design tool: Firstly, a fully automated procedure, where the prototype is created within the press of a button and secondly, it provides a process where the development team can add programming code to the prototype and thereby enrich its functionality.

The automatic process creates prototypes as quickly and easily as possible. The functionality of prototypes generated in this fashion is, however, limited. It does not support more elaborated prototypes than *click dummies*: prototypes with static interface screen changes that are triggered by a button press or gesture listener. However, click dummies are a helpful and commonly used tool in early design stages where the speed to produce a first testable approach is more important than an elaborated prototype behavior. Many commercial approaches discussed in Sect. 2.2 exclusively focus on click dummies.

Additionally, the automated mechanisms give the design team an opportunity to already experience their freshly designed ideas in the prototype during the design process. A designer can quickly test and adjust a draft without being pulled out of the collaborative ideation process for too long.

However, as the prototype evolves, the questions that are relevant to the design team will change. In early design stages prototypes are usually simple demo cases that provide a general feedback. Later on, more specific questions about certain prototype aspects or experiences from long-term usage tests will become relevant. For such prototypes, a more elaborate functionality than those of click dummies is necessary. This is why "Blended Prototyping" supports processes for enriching the functionality of a prototype by adding programming code. Developers can use code written in the programming language Java. Java is the language of the most popular operating system for mobile devices, Android. "Blended Prototyping" offers a process that compiles and integrates the added programming code dynamically into the app prototype so that it is executed in the app as native programming code. A more detailed description of this process, and of mechanisms we implemented for an easy code editing, can be found in Bähr (2013).

Technologies for the design of early prototypes usually base the implemented functionality on web-based technologies or pseudo code (compare Sect. 2.2). This allows them to be distributed to different mobile operating systems, at the cost of the capabilities of the programming code. When it comes to the programming of the later product, prototype code written in the native language can be integrated far more easily. The prospect of having to write code again from scratch naturally discourages developers to put much effort in the prototype development. Furthermore, native programming codes can use the same programming libraries and gets granted the same access rights as the later software.

Module 3—Testing Tool

The "Blended Prototyping" framework includes a testing tool that helps to evaluate prototypes in user tests. It provides a client that inflates and runs the prototype on a mobile device, it offers a server that is used to plan and conduct studies, and it includes logging mechanisms that build the basis to analyze a study.

Test users that want to participate in a study have to install an app only once. This app will then dynamically download and execute prototype data, without the user's further notice. Therefore, designers are free to alter their prototypes as they like and do not have to ask the test users to install a new version of the app prototype.

The testing tool supplies a Java server where prototypes are stored and distributed. For user management, the server uses a regular CSV file where user credentials are generated and the study design is planned. When a client addresses the server it checks the access rights provided to the delivered credentials and delivers the according prototype to the mobile device.

While a user tests the prototype, the client app generates logging data that is then uploaded to the server. Depending on its nature, each user control defined for the interface tracks data differently. All misses, touches that were registered by the screen and could not be addressed to a specific user control, are tracked as well.

The logging data is then uploaded to the server, where it is stored in a database that can be used by the development team for the test analysis. The data can be processed into a video where the user's interaction with the interface is displayed. More specific questions can be answered as well, such as the average time users stay on certain pages, for example.

4.3 Expert Reviews and Requirements Analysis

4.3.1 Motivation for New KPIs

As outlined above, user interface prototypes are very important for exploring and testing ideas early enough to be still able to consider changes in the development phase. A number of prototyping approaches exist, as highlighted above in Sect. 2. Each of these approaches addresses specific development conditions and stages. However, their authors have different views on which mobile user interface designers requirements should be addressed and take precedence.

A comprehensive catalogue is missing, which outlines the most prominent requirements of developers and designers with respect to app prototyping tools.

Such a catalogue should be used to provide guidelines for the user-centered design of new prototyping techniques. This could be used as a basis for metrics for comparing different prototyping approaches with each other. Related works provide a fragmented picture of numerous aspects that are considered the most important requirements. A comprehensive taxonomy cannot be found.

To fill this gap, we identified a catalogue of requirements from a literature research that we then evaluated with invited industry experts. The evaluation had two main objectives: Firstly, to ask the experts about requirements we might have missed, and secondly, to allow the experts to rate the importance of individual requirements. Literature and praxis suggest that the relevance of requirements changes in the course of a project and over the development cycle. Therefore, we asked the experts to assess the requirements' importance at different development stages.

4.3.2 Identifying a Requirements Catalogue Through Literature Research

As a first step in developing a requirements catalogue for mobile user interface prototyping tools we conducted a literature research in which we systematically surveyed related articles covering the last ten years of research in the field of human computer interaction. The result of these efforts is summarized in the Table 1, which lists the requirement categories with a short description and exemplary references.

4.3.3 Assessment of the Requirements Catalogue with Experts

Study Objectives

The first objective of the study was to allow experts add items to the catalogue we might have missed in the literature. Moreover, we wanted to assess the relevance of the requirement for the design and development process, both specifically for different stages and generally throughout the whole process. From this data, we planned to identify the most important requirements for early, middle, and late development stages.

At the time we performed the expert talks, the "Blended Prototyping" was already implemented to an extent, where the experts were able to try out and explore the technique independently. Hence, we introduced the approach to experts and asked them to provide feedback in two ways, namely with open comments, as well as in a ranking of the system with the identified requirements catalogue. In this way, we gained new insights for the further system development and at the same time tested the applicability of the requirements dimensions to judge a prototyping technique.

Table 1 List of requirements from the literature study

Requirement	Description	Exemplar References
Freedom of design	The tool allows designing in a free manner. Questions on accuracy can be postponed. Free creative work is promoted	(Davis et al. 2007; de Sá et al. 2008; Segura and Barbosa 2013)
Getting quick prototypes	The tool is targeted at delivering testable prototypes as quickly as possible	(Cherubini et al. 2007; Landay 1996; Newman et al. 2000; Szekely 1994)
Independent parallel development of designs	The tool helps and motivates users, to work on design alternatives and consider them in the testing	(Cherubini et al. 2007; Landay and Myers 2009; Snyder 2003)
Collocated group work	The tool is well suited to support groups to work simultaneously at the same place	(Holzmann and Vogler 2012; Karin and André 2009; Newman et al. 2000)
Remote group work	The tool is well suited to support groups to work simultaneously at different places	(Borchers et al. 2002; Horst 2011; Hupfer et al. 2004)
Support of expert reviews	The tool is well suited to conduct expert reviews	(Cherubini et al. 2007; Landay 1996)
Support of design reviews	The tool is well suited to conduct design reviews	(Korhonen et al. 2009; Nielsen and Molich 1990)
Tests in the real use context	The tool produces prototypes that can be tested in the same use context, as the later product	(Davis et al. 2007; Monrad Nielsen et al. 2006; de Sá et al. 2008)
Easy setup and distribution of user tests	The tool supports in the setup and execution of user tests and supplies mechanisms to deliver the prototype to the test users	(Holzmann and Vogler 2012; Lumsden and MacLean 2008; Monrad Nielsen et al. 2006)
Simultaneous tests of different ideas	The tool allows a comparative test of different design ideas	(Holzmann and Vogler 2012; Lumsden and MacLean 2008; Monrad Nielsen et al. 2006)
Tests with large numbers of users	The tool allows tests with large numbers of users, since test users can do the testing autonomously	(Derboven et al. 2010; Monrad Nielsen et al. 2006)
Advanced functionality of the prototype	The tool allows to produce prototypes of an elaborated functionality	(Koivisto and Suomela 2007; Maloney et al. 2010; Axure 2014)
Reusable prototypes	The tool makes it possible to reuse the prototype in later stages	(Holzmann and Vogler 2012; Lumsden and MacLean 2008; Monrad Nielsen et al. 2006)
Reusable programming code	The tool uses programming code that can be reused later in the development	(Korhonen et al. 2009; Lumsden and MacLean 2008; Monrad Nielsen et al. 2006)
Tests on different platforms	The tool creates prototypes that can be tested across different platforms, without major adaptions	(Dalmasso et al. 2013; Smutny 2012)

(continued)

Table 1 (continued)

Requirement	Description	Exemplar References
Use of animations	The tool allows the use and definition of dynamic content and animations in the prototype test	(Koivisto and Suomela 2007; Korhonen et al. 2009)
Automated model based evaluation	The tool supplies mechanisms, which can be used for automated model based evaluations of the interface prototype	(Amant et al. 2007; Paterno 2000)

Participants
We had interviews with a total number of 15 experts from Berlin and Potsdam (5 female, 10 male), all of whom had a professional background in mobile app development and design for at least two years. The description of their role in the development process varied between software developer, designer, user researcher and project manager. Maybe unsurprisingly for the Berlin startup scene, about a half answered that they were active in more than one of these roles at the same time. The answer regarding the average number of users of the apps developed by the experts varied between 10 users and 50 million users.

Procedure and Collected Data
We invited one expert at a time. At the start of each session, we explained the general purpose of the study and asked for permissions for video and data recordings. The survey was then structured into the three phases explained below. Questionnaires were answered on the computer, allowing us to implement dynamic question items.

Pre-questionnaire

In the beginning, questions on demographics and personal experience in app development and design questions were asked. Followed by that, we gathered data about the usual development processes of the experts and about the tools they apply. Next, the requirements we identified in the literature review were introduced. The experts were asked to rate each requirement regarding its importance (5 point scale, ranging from 1, *extremely unimportant to 5, extremely important*) at five different project stages (5 point scale, ranging from *very early* to *very late*). Hence, five data points were generated for each requirement.

In the following we asked the experts to suggest additional requirements that we might have missed in the presented catalogue. These newly suggested requirements were then rated on the same scale as the entries of our catalogue.

Demo and Free Discussion

After the participants completed the pre-questionnaire, they were introduced to the "Blended Prototyping" platform. After explaining the fundamental ideas of the approach, the expert was guided through a participatory demo, which followed a fixed script. We explained the use of the system step by step, allowing the expert to control the setup herself. Questions and debates about the system were always allowed and promoted. After the demo a free discussion was initiated. To allow later analysis, the sessions were recorded in video and audio.

Post-questionnaire

After the free debate came to an end, a second questionnaire was processed. Here, each expert was asked to rate the "Blended Prototyping" platform with the categories from literature and the ones, which were suggested by the expert in the first questionnaire. Additionally, we asked to give free feedback in open text fields.

Results from Expert Talks

For the sake of brevity, the following results of the expert talks are outlined in an aggregated form. A more detailed discussion can be found in the corresponding article (Bähr 2015).

Unfortunately, new requirements proposed by the experts were limited: Only 3 out of the 15 experts gave us suggestions, and no more than five new attributes were proposed. The '*Usability of the tool itself*' was the only category named twice. All the other suggestions were given by just one expert: '*good tutorials and help*', '*fun to use*', '*cross platform tool use*', '*open source availability*', and '*extended functionality*'. Therefore, too little data was generated to evaluate these suggestions in a reasonable way.

The average rating of the categories from literature is displayed in the single plots in Fig. 4. The change of relevance over project phases is expressed, and the

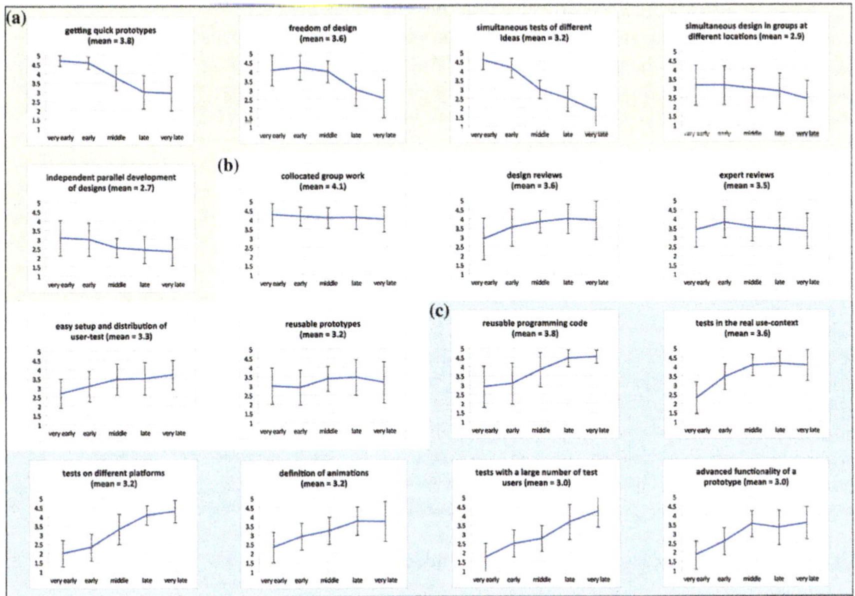

Fig. 4 Importance of the requirements at different development stages

confidence intervals for the means are displayed. Within the figure, requirements are grouped in a way that shows its changing importance at different project phases: (a) requirements that are important in the beginning, but less important in the end, (b) requirements that approximately keep their relevance throughout the development, and (c) requirements which are not very important in the beginning, but prove to be in the end. Within these groups, the requirements are sorted in descending order of their average importance over time. Due to an overall weak ranking, the category '*automated model-based evaluations*' was excluded from analysis.

The categories that were generally ranked highest are *collocated group work* (mean 4.1), *getting quick prototypes* (3.8) and *reusable programming code* (3.8). As the gradients of the plots in Fig. 4 indicate, we found that the importance of most requirements vary over time. A particularly high delta of lowest and highest value were identified for the *simultaneous tests of different ideas* (Δ 2.8), *tests with high numbers of users* (Δ 2.5), and *tests on different platforms* (Δ 2.3). In the Fig. 5, the ratings for the most important requirements we could identify for early, middle, and late stages are displayed. Examples for attributes that change particularly markedly during the development cycle are highlighted with colors.

The experts gave a generally positive feedback on "Blended Prototyping". Discussions about the system were vivid and extensive. Rating the system on the pre-defined requirements was successful; none of the experts had problems applying the metrics. On a 1 to 10 point scale, the "Blended Prototyping" platform was best scored to support *collaborative work* and *getting quick prototypes* (both 9.07 in average), followed by *tests with a large number of users* (8.14). The fewest points were given for *tests on different platforms* (5.77), *advanced functionality of a prototype* (5.21), and the *definition of animations* (3.15).

In the open discussion, the experts expressed that they particularly liked the use of simple paper in the design process. They appreciated the speed with which the paper design could be processed into a prototype. Those experts who had a

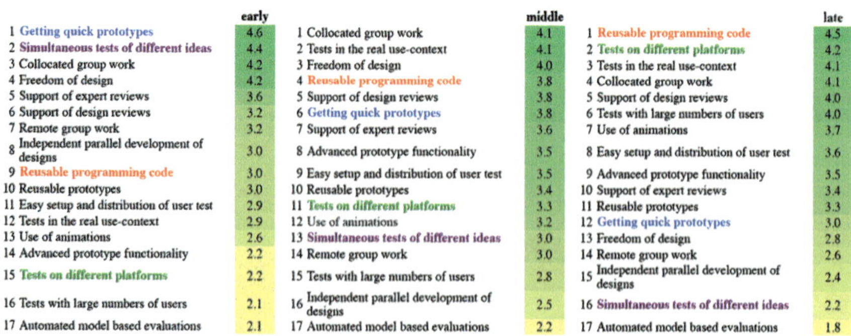

	early		middle		late
1 Getting quick prototypes	4.6	1 Collocated group work	4.1	1 Reusable programming code	4.5
2 Simultaneous tests of different ideas	4.4	2 Tests in the real use-context	4.1	2 Tests on different platforms	4.2
3 Collocated group work	4.2	3 Freedom of design	4.0	3 Tests in the real use-context	4.1
4 Freedom of design	4.2	4 Reusable programming code	3.8	4 Collocated group work	4.1
5 Support of expert reviews	3.6	5 Support of design reviews	3.8	5 Support of design reviews	4.0
6 Support of design reviews	3.2	6 Getting quick prototypes	3.8	6 Tests with large numbers of users	4.0
7 Remote group work	3.2	7 Support of expert reviews	3.6	7 Use of animations	3.7
8 Independent parallel development of designs	3.0	8 Advanced prototype functionality	3.5	8 Easy setup and distribution of user test	3.6
9 Reusable programming code	3.0	9 Easy setup and distribution of user test	3.5	9 Advanced prototype functionality	3.5
10 Reusable prototypes	3.0	10 Reusable prototypes	3.4	10 Support of expert reviews	3.4
11 Easy setup and distribution of user test	2.9	11 Tests on different platforms	3.3	11 Reusable prototypes	3.3
12 Tests in the real use-context	2.9	12 Use of animations	3.2	12 Getting quick prototypes	3.0
13 Use of animations	2.6	13 Simultaneous tests of different ideas	3.0	13 Freedom of design	2.8
14 Advanced prototype functionality	2.2	14 Remote group work	3.0	14 Remote group work	2.6
15 Tests on different platforms	2.2	15 Tests with large numbers of users	2.8	15 Independent parallel development of designs	2.4
16 Tests with large numbers of users	2.1	16 Independent parallel development of designs	2.5	16 Simultaneous tests of different ideas	2.2
17 Automated model based evaluations	2.1	17 Automated model based evaluations	2.2	17 Automated model based evaluations	1.8

Fig. 5 Most important requirements at early, middle, and late stages

professional background in programming saw the native programming approach as a significant advantage. A number of experts disliked the use of mobile devices in the design process for controlling the tabletop environment, criticizing the switch of media and negative effect of inter-group awareness. Some experts wished for better ways of implementing prototype functionality in the design process without entering programming code. At the same time, however, they expressed the concern that this could undermine the ease and reliability of the approach. Some of the experts specifically underlined that they liked the focus of the approach in concentrating on the design process with paper drawings, and explicitly neglecting more elaborate interactions with the system.

Discussion

The expert talks conducted were successful in identifying and assessing different sets of requirements for different development stages. Unfortunately, the consulted experts added few new requirements to our catalogue. The requirements presented by us were carefully determined in a literature review. However, we never expected, and still do not expect, the proposed catalogue to be complete.

Perhaps the question of naming additional requirements was too spontaneous? Maybe our guests were not motivated enough? Retrospective explanations are difficult to identify and evaluate. Possibly, expert group discussion would have improved the feedback. However, such discussions risk affecting experts' personal opinions, given that negative group effects such as the exclusion of single members, can occur.

In addition the identification of requirements, it was the purpose of the study to gain feedback on the "Blended Prototyping" approach and its implementation. We were excited to experience the fact that the invited experts felt interested enough about our work to stay much longer than we initially expected. On average, our guests left us after approximately three hours, during which a major share of the time was used for open feedback discussions. In a larger panel of experts, the extent of the individual feedback provided by the experts would have been reduced.

As Fig. 5 illustrates, the importance of requirements does change in the course of the development. Therefore, a Swiss Army Knife prototyping tool for all development conditions is hard to imagine. An approach should be more successful, if it concentrates on providing helpful mechanisms for particular development phases. The "Blended Prototyping" approach is targeted at early and middle design stages. Therefore, it should most importantly meet the requirements of *getting quick prototypes*, *simultaneous tests of different ideas*, *collocated group work*, *freedom of design*, and *tests in the real use-contexts*. The ratings provided by the experts on "Blended Prototyping" are particularly high in these dimensions. This result draws a promising picture of the potential of the "Blended Prototyping" approach. At the same time, it motivated us to evaluate the practical use of the approach within design teams, in comparison to other early design stage prototyping approaches.

4.4 Evaluation of "Blended Prototyping" in a User Study

When this text was written at the end of April 2015, the data analysis of the user study described in this section was not yet completed. Therefore, at this place, only limited first results will be displayed. A profound analysis and discussion of the matter will be submitted for publication at the ACM International Conference on Interactive Tabletops and Surfaces (ITS) 2015.

Study Objectives

The user study discussed here, evaluates "Blended Prototyping" in comparison to two other approaches for early design stage user interface prototyping: Paper Based Prototyping and computer based prototyping with the professional mockup software Axure. In the study, we asked design teams to use these tools to work on a given creative task. The study surveyed the ability of the different tools to meet the requirements we identified beforehand in expert reviews (see Sect. 4.3), and set this data in relationship to the overall quality of the created prototypes. In particular, we addressed the following questions to the study: How well are the single tools able to meet the requirements that are most important in early prototyping stages? Which aspects can be promoted best by the single approaches? Which prototyping tools are able to deliver the best results? To which degree does the tools scoring in the requirements determine the overall quality of the prototype results, as rated by designers in a post analysis?

Participants

A total number of 36 participants (17 female/19 male) were involved in the study. Subjects participated in groups of three, resulting in 12 groups. The participants were recruited from a subject database and blackboard postings in the context of two universities: the UdK Berlin and the TU Berlin.

Study Design

To get an invitation to the study, the participants first had to perform a short online survey, were basic programming knowledge and personal creativity were assessed. The groups were then built under the premise, to result in a comparable distribution of skills in each group. However, a complete balancing of the groups' potential is not possible. Therefore, we conducted the study in a within-subject design, meaning that each group did a creative task with each of the prototyping tools. The order in which the teams worked with the tools was balanced throughout the experiment. The selection of the design tasks was randomized, however, and one team never worked on the same task twice.

Each session was divided into to following five segments:

(a) Introduction to the tool, with a demo of the basic functionality
(b) Application of the tool in a short sample design task, to clear occurring questions and to demonstrate the basic understanding of the tool use
(c) Use of the tool in the actual design task (90 min)

(d) Presentation of the design result, either in a user test session, or in a prototype demo
(e) Individual answers to the questionnaires, described below.

All of these phases except for (c) where the tool was used in the productive design task were not limited in terms of time. Therefore, the total time needed for a session slightly varied from 3.5 to 4 h. Each group participated in three sessions and therefore spent a bit less than 12 h in total with the test. We therefore scheduled each test for 1.5 days on a weekend. This unusual high time commitment of the test subjects was rewarded with a comparatively high gratification of €120.

Collected Data

Questionnaires

At the end of each tool-use session, the subjects were asked to individually answer two questionnaires. The answers were related solely to the previously used tool and group work. First, the AttracDiff (Hassenzahl 2004; Hassenzahl et al. 2003) questionnaire was answered, which measures pragmatic quality, identification, stimulation and attractiveness of interactive products. Followed by that, the participants answered a questionnaire on different aspects of collaborative work and interpersonal relations, which was established by Sauppé and Mutlu (2014). In particular the second questionnaires included scales on rapport, teamwork, the ability to collaborate, empathic concern, perspective taking, interpersonal solidarity and homophily.

When the sessions on all three design-tools were completed, a third questionnaire concluded the test. Here, the participants were asked about their previous experience with all three presented prototyping tools.

Video Data and Created Prototypes

All phases of the design and presentation sessions were recorded on audio and video, so that a later analysis of the creative teamwork process and prototype outcome was made possible. In addition to that, the prototypes that resulted from the 36 creative sessions were put aside for the later analysis of the productive outcome. The form in which this prototype data was saved depended on the tool used: Axure prototypes were saved digitally, Paper Based prototypes in their physical paper from and Blended prototypes as both, paper and software. Other content that was created by the design teams such as notes or pre-versions of the prototype were kept for later analysis as well.

The prototype results, as well as the videos of their tests, will be presented to external design experts for a rating of the results' design, imaginativeness, and ability to generate insights for the prototyping process. As a measure for the overall success of a prototype, the design experts will be additionally asked to rate the prototypes regarding their potential to be followed up in a subsequent design process. In addition to that, the experts will be asked as to whether they would financially invest in a further development of the prototype.

Furthermore, the collaborative performance of the tools within the single design sessions will be exploited in video analysis. For this, we used an existing taxonomy (Gutwin and Greenberg 2000; Reilly et al. 2005; Wallace et al. 2009) that we developed into an evaluation record. Such measures include factors that measure communication, awareness, and coordination within the group.

First Results

At the time of writing this text, the data analysis had not been fully completed. Expert ratings of the collaborative performance based on videotapes of the experimental tasks are not completed yet, due to their time consuming character. Therefore, the results presented in this section discuss questionnaire data exclusively.

Bar charts above in Fig. 6 display the mean ratings of the three tools on the subscales of the AttrakDiff questionnaire. Upon visual inspection of the data it becomes apparent that mean ratings do not differ largely between the four tools: mean ratings vary around a value of four, the mean value of the scale itself. However, Axure was consistently rated lowest in all scales. "Blended Prototyping" was generally rated better than Paper Based Prototyping, especially on the subscales *hedonic* quality—stimulation and *attractiveness*.

The second group of bar charts in Fig. 7 shows the mean ratings of the tools on the subscales of the questionnaire on collaborative work. Here, the results draw an inconclusive picture. Again, Axure tends to be rated lower than the other tools. However, the differences between mean ratings are very small. They vary closely around the middle rating of 4. This tendency to the middle indicates that the used scales were not able to produce a distinctive rating by the test subjects.

To investigate the significance of the measured means discussed above, data was analyzed using one-way repeated measures analysis of variances (ANOVA) with the factor *tool* having three levels (the experimental groups "Blended Prototyping",

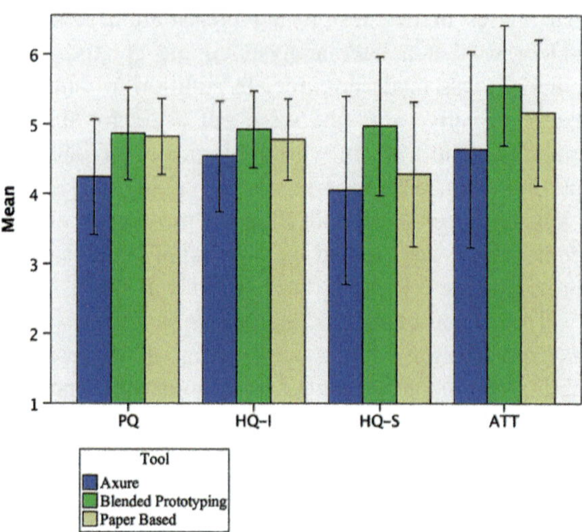

Fig. 6 Boxplots of means for AttrakDiff subscales *PQ* Pragmatic Quality, *HQ-I* Hedonic Quality Identification, *HQ-S* Hedonic Quality Stimulation, and *ATT* Attractiveness (Error Bars: ±SD)

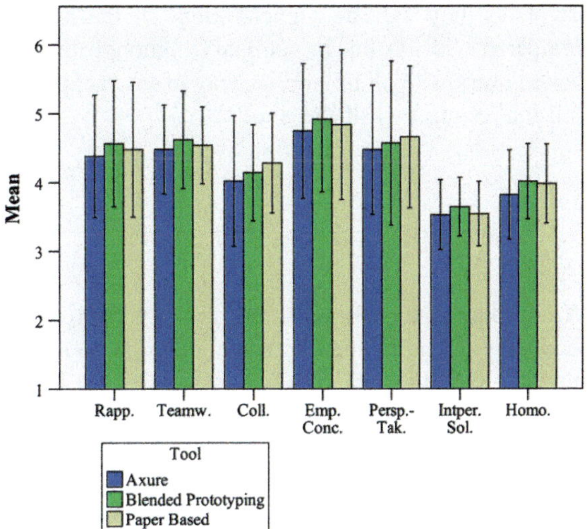

Fig. 7 Boxplots of means for Rapport, Teamwork, Collaborativeness, Empathic Concern, Perspective Taking, Interpersonal Solidarity, and Homophily (Error Bars: ±SD)

Axure, and Paper Based Prototyping). All questionnaire subscales were analyzed separately.

The results of the questionnaire data regarding collaborative work did not prove to be statistically significant. However, repeated measures ANOVA revealed a main effect for the factor *tool* on three of the four AttracDiff subscales.

For the subscales *pragmatic quality* (F (1.708, 59.765) = 9.796, p = 0.000, η^2_{part} = 0.219, Greenhouse-Geisser corrected) and *hedonic quality—stimulation* (F (2, 70) = 9.755, p = 0.000, η^2_{part} = 0.218) post hoc Bonferroni comparisons revealed significant differences between "Blended Prototyping" and Axure, as well as between Axure and Paper Based Prototyping. The difference between Paper Based Prototyping and "Blended Prototyping" did not prove to be significant.

For the subscale *attractiveness* (F (1.683, 58.922) = 6.657, p = 0.004, η^2_{part} = 0.160), significant differences were found only between "Blended Prototyping" and Axure. Differences between the other tools did not prove to be significant.

Furthermore, due to different ratings for "Blended Prototyping" and Axure, a statistical trend (F (1.52, 53.467) = 2.973, p = 0.073, η^2_{part} = 0.078) was observed for the subscale *hedonic quality—identification*.

The data collected from the questionnaire therefore does not allow statements on the success of the different tools, i.e. their ability to facilitate collaborative work. Here, the objective investigation from video analysis of the level of creative teamwork is necessary. However, the AttracDiff was able to show statistical distinctions between the rated pragmatic quality, hedonic quality stimulation and attractiveness. These results indicate a higher perceived usability and overall attractiveness of the "Blended Prototyping" tool when compared to classical desktop prototyping software like Axure. Furthermore, users seem to appreciate the functionality of the "Blended Prototyping" tool and its innovativeness and interest factor, as they indicate higher ratings of the hedonic quality identification when compared with Axure. In summary, although the "Blended Prototyping" tool presented users with an unfamiliar way of interaction, it appears to be the most usable and interesting tool of the three.

5 Discussion and Further Work

We developed the "Blended Prototyping" approach in a process based on iterative feedback and refinements. Three different tools were developed that make the thoughts of "Blended Prototyping" applicable in productive design and development sessions. The platform was successfully tested in interviews with domain experts and in user-tests with experienced students.

The research on "Blended Prototyping" took a holistic approach that included interdisciplinary research questions. It delivered a number of results that can prove relevant to different research fields. In the following, a number of fields where we see the most prominent contribution will be highlighted and our personal next research steps are portrayed.

As displayed in Sect. 4.3, we conducted a literature study to identify requirements for mobile UI prototyping tools, which we then evaluated with experts. Such a list of requirements can serve as a reference and design guideline to develop prototyping tools. Further research is needed to further contribute to this catalogue and to adapt it to changes that might occur to the development conditions for mobile app designers and developers. The results from the user study, discussed in Sect. 4.4, have not yet been fully analyzed. First insights we gained from questionnaire data draw a promising picture of "Blended Prototyping's" capabilities of proving to be a usable, innovative and attractive collaborative tool.

Another interesting perspective from the area of computer science is the process of "Blended Prototyping" to produce, distribute and run programming code. In our approach, we implemented mechanisms that help to program native Android code in a focused pure Java fashion, which ignores a number of Android specific programming aspects. This makes the approach easier to understand, even for novice programmers with no experience in Android development. However, these Android specific concepts are very valuable to the later app, so at a certain stage they should be considered in the development. Questions as to how the "Blended Prototyping" approach affects the code quality, or how long the approach can be reasonably helpful in the development stages, are also of significant interest.

The "Blended Prototyping" design tool is currently based on a tabletop computing system. As outlined in Sect. 4.2, this assists when using physical paper as a center of the design. At the same time, however, it implements a complex hardware setup, which is not easily accessible. There is still a lot of potential to develop and test new interaction concepts for improving the human-computer-paper interaction space. Sketch analysis and recognition algorithms, or the inclusion of gestural definition concepts, could be helpful for developing approaches as alternatives to the tabletop computing setup. In this area, the focus should be on the intuitiveness and reliability of the interactions so as to avoid interfering with the free flow of team creativity.

We plan on making the "Blended Prototyping" platform open source so that others can use the system and contribute to the ongoing development. As displayed in Sect. 4.2, we structured the platform into independent modules that are linked to each other with open file standards. This should make it easy for others to develop and integrate new modules into the platform, or to develop linkages to other existing techniques.

We currently plan a follow-up research project that will extend the "Blended Prototyping" platform in a way that makes it applicable for the prototyping of Internet of Things (IoT) applications. The term IoT describes the phenomenon that more and more things become connected devices. This does not only relate to the old story of your fridge, automatically ordering a refill of milk over the internet. Gartner estimates, that the number of devices connected to the internet will grow to more than 25 billion by 2020, which is more than five times the number as is the case today.[1]

Recently, different hardware kits[2] have hit the market that assist amateur users in developing sensor-based hardware applications. Such kits usually use sensors that can be connected to programmable micro controller chipsets. The controllers can then be programmed in a supplied software environment.

At this stage, the prototyping kits primarily address programmers as their main target group. The supplied sensors are limited and do not fit well into use contexts. This makes it difficult for non-expert users to understand their meaning and application domains. Furthermore, current development tools are focused on computer software, which makes it difficult to use them in a collaborative context.

We plan on using the insights we gained from the "Blended Prototyping" research to develop a collaborative IoT prototyping tool. Here, ideas for IoT applications could be developed alongside with the mobile app in a tabletop computing environment that promotes collaborative work processes. Furthermore, we plan to conduct further research on how the complex field of sensory measurements could be simplified for novice use. What sensors are actually most needed? How can the scales of such sensors be simply communicated? What virtual sensors could be developed, and which aggregate complex measures into values that are easy to understand?

Our goal in this context is a platform that not only produces testable software user interfaces, but testable hardware prototypes as well. The "Blended Prototyping" platform is a good starting point for tackling this and other challenges concerning creative collaborative work.

---

[1]http://www.gartner.com/newsroom/id/2905717.

[2]e.g.: littleBits, WunderBar, spark.io.

Acknowledgments In the first year of the project, Stephanie Neumann supported "Blended Prototyping" as a designer, interaction-designer and photographer. Stephanie worked in the Design Research Lab of the UdK Berlin, lead by Prof. Dr. Gesche Joost. We want to express our gratitude for the thoughts and ideas the project gained from this collaboration. Furthermore, we want to thank Prof. Michael Rohs of the Leibniz University Hannover. He supported and advised the project steadily right from the start.

References

Amant, R. S., Horton, T. E., & Ritter, F. E. (2007). Model-based evaluation of expert cell phone menu interaction. *Transactions on Computer-Human Interaction, 14*(1), doi:10.1145/1229855.1229856

Anoto. (2014). *Digital writing technology.* Retrieved 3 June 2014 from http://www.anoto.com/

Axure. (2014). *Interactive wireframe software & mockup tool.* Retrieved June 3, 2014 from http://www.axure.com/

Bähr, B. (2013). Rapid creation of sketch-based native android prototypes with "blended prototyping". In *Mobile HCI 2013—Workshop on Prototyping to Support the Interaction Designing in Mobile Application Development*, München.

Bähr, B. (2015). Towards a requirements catalogue for prototyping tools of mobile user interfaces. In *Proceedings of the 9th HCI International Conference,* (Los Angeles, 2015), in press.

Bähr, B., & Neumann, S. (2013). "Blended prototyping" design for mobile applications. In *Rethinking Prototyping: Proceedings of the Design Modelling Symposium* (pp. 68–80), epubli, Berlin.

Bailey, B. P., Biehl, J. T., Cook, D. J., & Metcalf, H. E. (2008). Adapting paper prototyping for designing user interfaces for multiple display environments. *Personal Ubiquitous Computing, 12*(3), 269–277. doi:10.1007/s00779-007-0147-2

Balsamiq. (2014) *Rapid, effective and fun wireframing software.* Retrieved June 3, 2014 from http://www.balsamiq.com

Battocchi, A., Pianesi, F., Tomasini, D. et al. (2009). Collaborative puzzle game: A tabletop interactive game for fostering collaboration in children with autism spectrum disorders (ASD). In *Proceedings of the ACM International Conference on Interactive Tabletops and Surfaces* (pp. 197–204). New York: ACM. doi:10.1145/1731903.1731940

Baum, L. F., & Denslow, W. W. (2014). *The wonderful wizard of oz.* North Charleston: CreateSpace Independent Publishing Platform.

Beyer, H., & Holtzblatt, K. (1997). *Contextual design: Defining customer-centered systems (interactive technologies).* San Diego: Academic Press.

Borchers, J., Ringel, M., Tyler, J., & Fox, A. (2002). Stanford interactive workspaces: A framework for physical and graphical user interface prototyping. *IEEE Wireless Communications, 9*(6), 64–69. doi:10.1109/MWC.2002.1160083

Brewster, S. (2002). Overcoming the lack of screen space on mobile computers. *Personal Ubiquitous Computing, 6*(3), 188–205.

Cherubini, M., Venolia, G., DeLine, R., & Ko, A. J. (2007) Let's go to the whiteboard: How and why software developers use drawings. In *Proceedings of the SIGCHI Conference on Human Factors in Computing Systems* (pp. 557–566). New York: ACM. doi:10.1145/1240624.1240714

Consolvo, S., & Walker, M. (2003). Using the experience sampling method to evaluate ubicomp applications. *IEEE Pervasive Computing, 2*(2), 24–31.

Cook, D. J., & Bailey, B. P. (2005). Designers' use of paper and the implications for informal tools. In *Proceedings of the 17th Australia conference on Computer-Human Interaction: Citizens Online: Considerations for Today and the Future* (pp. 1–10). New York: ACM. Retrieved June 2, 2014 from http://dl.acm.org/citation.cfm?id=1108368.1108402

Coyette, A., Kieffer, S., & Vanderdonckt, J. (2007). Multi-fidelity prototyping of user interfaces. In C. Baranauskas, P. Palanque, J. Abascal & S. D. Junqueira Barbosa (Eds.), *Human-computer interaction—INTERACT 2007* (pp. 150–164). Berlin: Springer. Retrieved September 4, 2013 from http://link.springer.com/chapter/10.1007/978-3-540-74796-3_16

Dalmasso, I., Datta, S. K., Bonnet, C. & Nikaein, N. (2013) Survey, comparison and evaluation of cross platform mobile application development tools. In *Wireless Communications and Mobile Computing Conference* (pp. 323–328). Sardinia: IEEE. doi:10.1109/IWCMC.2013.6583580

Davis, R. C., Saponas, T.S., Shilman, M., & Landay, J. A. (2007). SketchWizard: Wizard of oz prototyping of pen-based user interfaces. In *Proceedings of the 20th Annual ACM Symposium on User Interface Software and Technology* (pp. 119–128). New York: ACM. doi:10.1145/1294211.1294233

de Sá, M., & Carriço, L. (2006). Low-fi prototyping for mobile devices. In *CHI'06 Extended Abstracts on Human Factors in Computing Systems* (pp. 694–699). New York: ACM. doi:10.1145/1125451.1125592

de Sá, M., & Carriço, L. (2008a). Defining scenarios for mobile design and evaluation. In *CHI'08 Extended Abstracts on Human Factors in Computing Systems* (pp. 2847–2852). New York: ACM. doi:10.1145/1358628.1358772

de Sá, M., & Carriço, L. (2008b). Lessons from early stages design of mobile applications. In *Proceedings of the 10th International Conference on Human Computer Interaction with Mobile Devices and Services* (pp. 127–136). New York: ACM. doi:10.1145/1409240.1409255

de Sá, M., Carriço, L., Duarte, L., & Reis, T. (2008). A mixed-fidelity prototyping tool for mobile devices. In *Proceedings of the Working Conference on Advanced Visual Interfaces* (pp. 225–232). New York: ACM. doi:10.1145/1385569.1385606

Derboven, J., De Roeck, D., Verstraete, M., Geerts, D., Schneider-Barnes, J., & Luyten, K. (2010). Comparing user interaction with low and high fidelity prototypes of tabletop surfaces. In *Proceedings of the 6th Nordic Conference on Human-Computer Interaction: Extending Boundaries* (pp. 148–157). New York: ACM. doi:10.1145/1868914.1868935

Duh, H., Tan, G., & Chen, V. (2006) Usability evaluation for mobile device: A comparison of laboratory and field tests. In *Proceedings of the 8th Conference on Human-Computer Interaction with Mobile Devices and Services* (pp. 181–186). New York: ACM Press. Retrieved from doi:10.1145/1152215.1152254

Gutwin, C., & Greenberg, S. (2000). The mechanics of collaboration: Developing low cost usability evaluation methods for shared workspaces. In *Proceedings of the 9th IEEE International Workshops on Enabling Technologies: Infrastructure for Collaborative Enterprises* (pp. 98–103). Gaithersburg: IEEE Computer Society. Retrieved 30 April 2015 from http://dl.acm.org/citation.cfm?id=647068.715651

Hassenzahl, M. (2004). The interplay of beauty, goodness and usability in interactive products. *Human-Computer Interaction, 19*(4), 319–349. doi:10.1207/s15327051hci1904_2

Hassenzahl, M., Burmester, M., & Koller, F. (2003). AttrakDiff: Ein Fragebogen zur Messung wahrgenommener hedonischer und pragmatischer Qualität. *Mensch & Computer 2003, 57*, 187–196. Retrieved May 18, 2015 from http://link.springer.com/chapter/10.1007/978-3-322-80058-9_19

Holzmann, C., & Vogler, M. (2012). Building interactive prototypes of mobile user interfaces with a digital pen. In *Proceedings of the 10th Asia Pacific Conference on Computer Human Interaction* (pp. 159–168). New York: ACM. doi:10.1145/2350046.2350080

Horst, W. (2011). Supportive tools for collaborative prototyping. Nordes 0, 4. Retrieved June 2, 2014 from http://www.nordes.org/opj/index.php/n13/article/view/147

Hupfer, S., Cheng, L. T., Ross, S., & Patterson, J. (2004). Introducing collaboration into an application development environment. In *Proceedings of the 2004 ACM Conference on Computer Supported Cooperative Work* (pp. 21–24). New York: ACM. doi:10.1145/1031607.1031611

James, L., Newman, M. W., Hong, J. I., & Landay, J. A. (2000). DENIM: Finding a tighter fit between tools and practice for web site design. In *Proceedings of the SIGCHI Conference on Human Factors in Computing Systems* (pp. 510–517). New York: ACM. doi:10.1145/332040.332486

Karin, L., & André, E. (2009). MoPeDT: Features and evaluation of a user-centred prototyping tool. In *Proceedings of the 2nd ACM SIGCHI Symposium on Engineering Interactive Computing Systems* (pp. 93–102). New York: ACM. doi;10.1145/1822018.1822033.

Kieffer, S., Coyette, A., & Vanderdonckt, J. (2010). User interface design by sketching: A complexity analysis of widget representations. In *Proceedings of the 2nd ACM Sigchi Symposium on Engineering Interactive Computing Systems* (pp. 57–66). New York: ACM. doi:10.1145/1822018.1822029

Kjeldskov, J., Skov, M. B., Als, B. S., & Høegh, R. T. (2004) Is it worth the hassle? Exploring the Added value of evaluating the usability of context-aware mobile systems in the field. In *Mobile Human-Computer Interaction—MobileHCI 2004* (pp. 529–535). Berlin: Springer. doi:10.1007/978-3-540-28637-0_6

Klemmer, S. R., Everitt, K. M., & Landay, J. A. (2008). Integrating physical and digital interactions on walls for fluid design collaboration. *Human-Computer Interaction, 23*(2), 138–213. doi:10.1080/07370020802016399

Klemmer, S. R., Newman, M. W., Farrell, R., Bilezikjian, M., & Landay, J. A. (2001). The designers' outpost: A tangible interface for collaborative web site. In *Proceedings of the 14th Annual ACM Symposium on User Interface Software and Technology* (pp. 1–10). New York: ACM. doi:10.1145/502348.502350

Klemmer, S. R., Sinha, A. K., Chen, J., Landay, J. A., Aboobaker, N., & Wang, A. (2000). Suede: A wizard of oz prototyping tool for speech user interfaces. In *Proceedings of the 13th Annual ACM Symposium on User Interface Software and Technology* (pp. 1–10). New York: ACM. doi:10.1145/354401.354406

Koivisto, E. M. I., & Suomela, R. (2007). Using prototypes in early pervasive game development. In *Proceedings of the SIGGRAPH Symposium on Video Games* (pp. 149–156). New York: ACM. doi:10.1145/1274940.1274969

Korhonen, H., Paavilainen, J., & Saarenpää, H. (2009). Expert review method in game evaluations: Comparison of two playability heuristic sets. In *Proceedings of the 13th International MindTrek Conference: Everyday Life in the Ubiquitous Era* (pp. 74–81). New York: ACM. doi:10.1145/1621841.1621856

Landay, J. A. (1996). SILK: Sketching interfaces like krazy. In *Conference Companion on Human Factors in Computing Systems* (pp. 398–399). New York: ACM. doi:10.1145/257089.257396

Landay, J. A., & Myers, B. A. (2009). Just draw it! Programming by Sketching Storyboards.

Lumsden, J., & MacLean, R. (2008). A comparison of pseudo-paper and paper prototyping methods for mobile evaluations. In R. Meersman, Z. Tari & P. Herrero (Eds.), *On the move to meaningful internet systems: OTM 2008 Workshops* (pp. 538–547). Heidelberg: Springer. Retrieved July 23, 2013 from http://link.springer.com/chapter/10.1007/978-3-540-88875-8_77

Maloney, J., Resnick, M., Rusk, N., Silverman, B., & Eastmond, E. (2010). The scratch programming language and environment. *ACM Transaction on Computing Education, 10*(4), 1–15.

McAdam, C., & Brewster, S. (2011) Using mobile phones to interact with tabletop computers. In *Proceedings of the ACM International Conference on Interactive Tabletops and Surfaces* (pp. 232–241). New York: ACM. doi:10.1145/2076354.2076395

MockFlow. (2014). *Super-easy wireframing*. Retrieved June 3, 2014 from http://www.mockflow.com/

Monrad Nielsen, C., Overgaard, M., Bach Pedersen, M., Stage, J., & Stenild, S. (2006). It's worth the hassle!: The added value of evaluating the usability of mobile systems in the field. In *NordiCHI'06* (pp. 272–280). New york: ACM.

Newman, M. W., & Landay, J. A. (2000). Sitemaps, storyboards, and specifications: A sketch of web site design practice. In *Proceedings of the 3rd Conference on Designing Interactive Systems: Processes, Practices, Methods, and Techniques* (pp. 263–274). New York: ACM.

Newman, M. W., Lin, J., Hong, J. I., & Landay, J. A. (2003). DENIM: an informal web site design tool inspired by observations of practice. *International Journal of Human-Computer Interaction, 18*(3), 259–324.

Nielsen, J. (1993). Iterative user-interface design. *Computer, 26*(11), 32–41.

Nielsen, J., & Molich, R. (1990). Heuristic evaluation of user interfaces. In *Proceedings of the SIGCHI Conference on Human Factors in Computing Systems* (pp. 249–256). New York: ACM. doi:10.1145/97243.97281

Paterno, F. (2000). *Model-based design and evaluation of interactive applications*. Heidelberg: Springer.

Piper, B., Ratti, C., & Ishii, H. (2002). Illuminating clay: A 3-D tangible interface for landscape analysis. In *Proceedings of the SIGCHI Conference on Human factors in Computing Systems: Changing Our World, Changing Ourselves* (pp. 355–362). New York: ACM.

Radatz, J. (1997). *The IEEE standard dictionary of electrical and electronics terms*. New York: IEEE Standards Office.

Rädle, R., Jetter, H. C., Marquardt, N., Reiterer, H., & Rogers, Y. (2014). HuddleLamp: Spatially-aware mobile displays for Ad-hoc around-the-table collaboration. In *Proceedings of the Ninth ACM International Conference on Interactive Tabletops and Surfaces* (pp. 45–54). New York: ACM. doi:10.1145/2669485.2669500

Reilly, D., Dearman, D., Welsman-Dinelle, M., & Inkpen, K. (2005). Evaluating early prototypes in context: trade-offs, challenges, and successes. *IEEE Pervasive Computing, 4*(4), 42–50. doi:10.1109/MPRV.2005.76

Ringel Morris, M., Paepcke, A., & Winograd, T. (2006). TeamSearch: Comparing techniques for co-present collaborative search of digital media. In *Proceedings of the First IEEE International Workshop on Horizontal Interactive Human-Computer Systems*, IEEE Computer Society (pp. 97–104).

Sauppé, A., & Mutlu, B. How social cues shape task coordination and communication. In *Proceedings of the 17th ACM Conference on Computer Supported Cooperative Work & Social Computing* (pp. 97–108). New York: ACM. doi:10.1145/2531602.2531610

Schumann, J., Strothotte, T., Laser, S., & Raab, A. (1996). Assessing the effect of non-photorealistic rendered images in CAD. in *Proceedings of the SIGCHI conference on Human factors in computing systems: common ground* (pp. 35–41). New York: ACM. doi:10.1145/238386.238398

Segura, V. C. V. B., & Barbosa, S. D. J. (2013). UISKEI ++: Multi-device wizard of oz prototyping. In *Proceedings of the 5th ACM SIGCHI Symposium on Engineering Interactive Computing Systems* (pp. 171–174). New York: ACM. doi:10.1145/2480296.2480337

Seifert, J., Pfleging, B., del Carmen, E., Bahamóndez, V., Hermes, M., Rukzio, E. and Schmidt, A. (2011). Mobidev: A tool for creating apps on mobile phones. In *Proceedings of the 13th International Conference on Human Computer Interaction with Mobile Devices and Services* (pp. 109–112). New York: ACM. doi:10.1145/2037373.2037392

Shen, C., Lesh, N. B., Vernier, F., Forlines, C., & Frost, J. (2002). Sharing and building digital group histories. In *Proceedings of the 2002 ACM Conference on Computer Supported Cooperative Work* (pp. 324–333). New York: ACM.

Smutny, P. (2012). Mobile development tools and cross-platform solutions. In *Carpathian Control Conference (ICCC), 2012 13th International* (pp. 653–656). doi:10.1109/CarpathianCC.2012.6228727

Snyder, C. (2003). *Paper prototyping: The fast and easy way to design and refine user interfaces (the morgan kaufmann series in interactive technologies)*. San Francisco: Morgan Kaufmann Publishers.

Spindler, M., Stellmach, S., & Dachselt, R. (2009). PaperLens: Advanced magic lens interaction above the tabletop. In *Proceedings of the ACM International Conference on Interactive Tabletops and Surfaces* (pp. 69–76). New York: ACM.

Szekely, P. (1994). User interface prototyping: Tools and techniques. In *ICSE Workshop on SE-HCI* (pp. 76–92). Retrieved from http://citeseerx.ist.psu.edu/viewdoc/summary?doi=10.1.1.41.6764

Tuddenham, P., Davies, I., & Robinson, P. (2009). WebSurface: An interface for co-located collaborative information gathering. In *Proceedings of the ACM International Conference on Interactive Tabletops and Surfaces* (pp. 181–188). New York: ACM.

Underkoffler, J., & Ishii, H. (1999). Urp: A luminous-tangible workbench for urban planning and design. In *Proceedings of the SIGCHI conference on Human Factors in Computing Systems: The CHI is the Limit* (pp. 386–393). New York: ACM.

Wallace, J. R., Scott, S. D., Stutz, T., Enns, T., & Inkpen, K. (2009). Investigating teamwork and taskwork in single- and multi-display groupware systems. *Personal Ubiquitous Computing, 13* (8), 569–581. doi:10.1007/s00779-009-0241-8

Wellner, P. (1993). Interacting with paper on the DigitalDesk. *Communication of the ACM, 36*(7), 87–96.

Winters, F. J., Mielenz, C., & Hellestrand, G. (2004). Design process changes enabling rapid development. Retrieved 12 March 2014 from http://www.academia.edu/1266533/Design_process_changes_enabling_rapid_development

Zufferey, G., Jermann, P., Lucchi, A., & Dillenbourg, P. (2009). TinkerSheets: using paper forms to control and visualize tangible simulations. In *Proceedings of the 3rd International Conference on Tangible and Embedded Interaction* (pp. 377–384). New York: ACM.

Beyond Prototyping

Jussi Ängeslevä, Iohanna Nicenboim, Jens Wunderling and David Lindlbauer

Abstract "Beyond Prototyping" is a research undertaking exploring the possibilities of algorithmically defined products that can be easily manufactured using digital fabrication techniques. Using interdisciplinary teaching between two universities and collaborations with small commercial studios as well as a series of product-service systems to evaluate the feasibility and appeal of such products, beyond prototyping proposes a vision of service model that sits between atelier and mass production. *Locatable*, *Ciphering* and *Highlight*, three case studies implemented as web and material services, chronicle the challenges and opportunities of such products. To evaluate their success, the services are offered to the public, who are subsequently sent surveys to reflect on the products. The case studies demonstrate that such products have potential to complement the current market with new business models.

J. Ängeslevä (✉)
Institute for Time Based Media, Berlin University of the Arts,
Berlin, Germany
e-mail: jussi.angesleva@iki.fi

I. Nicenboim
Institute for Time Based Media, Berlin University of the Arts, Berlin, Germany
e-mail: iohanna.nicenboim@gmail.com

J. Wunderling
Institute for Industrial Design, Interaction Design Studies, HS Magdeburg-Stendal,
Stendal, Germany
e-mail: jens@jenswunderling.com

D. Lindlbauer
Computer Graphics, Technische Universität Berlin, Berlin, Germany
e-mail: david.lindlbauer@tu-berlin.de

C. Gengnagel et al. (eds.), *Rethink! Prototyping*,
DOI 10.1007/978-3-319-24439-6_10

1 Motivation

The research undertaking "Beyond Prototyping" has its roots in a series of collaborations, and transdisciplinary activities in the Digital Media Design department of the Berlin University of the Arts (UdK Berlin) and Computer Graphics department at the Technische Universität Berlin (TU Berlin). Perhaps due to the inherently hybrid approach of digital media design, as a discipline that combines visual communications, electronics, computer science, product design, human factors and many other perspectives through embodied activity, we had already conducted collaborative experiments between the departments years prior to the establishment of the "Rethinking Prototyping" research project.

For example, a course offering "Computational Photography" was a course we jointly conducted, wherein the visual communications opportunities of photography was contrasted with the general transformation in society in which every mobile device was suddenly a camera, and where computers could increasingly better "see" pictures. These kinds of collaborations were always carefully crafted to have meaningful challenges and opportunities fitting the syllabi of an art and design programme as well as a computer science degree.

Digital media design, or with the new name of "New Media Studio" is a course, that is exploring the communication potential of new technological means, and questioning their role and impact in the culture. This questioning is best done by actively using and exploring the enabling capabilities, in a hands on fashion, prototyping new experiences, new services or criticizing the status quo, as well as showing warning examples of what might happen if the worst sides of the explored technology were to become the standard.

Also, leading to the "Beyond Prototyping" undertaking were a series of explorations in digital fabrication that occurred in the years prior, such as a course "Beauty of Data", which looked at the translation of "machine readable internet", or "Web 2.0", which examined physical artefacts with fabrication technologies. Suddenly, when the friends of social networks were turned to physical sculptures on a mantelpiece, or a mobile network's potential for tracking one's movements was visualised as a 3D volume, the new tangibility of all this, triggered a different kind of interest in the otherwise invisible digital aura we carry around us. Similarly, a course entitled "Indie Design" looked at how digital fabrication can enable an explosion of independent design labels that do not rely on large-scale production facilities such as a factory, and how new kinds of individualisation can be enabled by algorithmically defined aesthetics that are easily configured by end-users.

In summary, "Beyond Prototyping" stemmed from a long lasting collaboration and exploration of a field that would combine technological and design challenges in equal proportions that the Digital Media Design and Computer Graphics departments would be able to fulfil and realise, together.

2 Teaching

The previous work leading to the formulation of the "Beyond Prototyping" project used university teaching at masters level as a testing ground for broadly scoping out the possibilities of an emerging field. A semester theme for a studio class would be intensively explored in weekly seminars, discussions and hands on building of prototypes and scenarios. The advantage of this method was the open-endedness and broad scope, where within a relatively short period of time one can see a landscape of interesting complementary or contrasting areas that follow a common theme. These explorations, even though they manifest themselves in individual concrete design projects, highlight a way of thinking and propose a way of thinking about a subject. But due to the time constraints (basically extending only over a single semester at best), the depth and level of execution often stays at the level that could be much improved with more time.

Unfortunately, this time expenditure cannot be justified under the normal teaching syllabus. Some ideas get then developed further in the students' individual work, or in their graduation projects, but they hardly extend to transdisciplinary collaborations.

We developed a strategy, informed by previous collaborations, as well as the theme-driven studio works, where we would start with a broad scoping through class teaching. The results would stay quite open-ended, but then in parallel develop a series of in-depth projects that would guide the whole research project time frame of three years, becoming much more refined, real world services, going beyond prototyping.

Exhibition in the Cloud The first transdisciplinary teaching project we conducted, (which already started before the official launch date of the "Rethinking Prototyping" research project) was a collaboration between New York and Berlin:

> *Prototype: Exhibition in the Cloud* is an interdisciplinary collaboration between Parsons The New School for Design and the Berlin University of the Arts (UdK Berlin) which seeks to challenge and reinvent received notions of prototyping, extending its design and industrial origin to encompass artistic imagination.
>
> Invoking the image of cloud computing, "prototype" here means a state of constant transformation and becoming. Like a cloud, it is amorphous and malleable, unstable and precarious. Instead of achieving a functionalist goal-oriented objective, prototyping-in-the-cloud becomes a mechanism of repetition in difference; always self-renewing and regenerating, revealing its infinite potential through chance, adaptivity and ephemerality in materiality. To prototype therefore is to invent the unforeseeable, to cast a shape that is at the same time formless. To prototype is to imagine the ineffable and to create polymorphic manifestations that are at once crystallized and fleeting. It is as much a way of cultural intervention as a mode of formal exercise in which memories, histories, locations and relations are engendered, tested, reiterated and distributed—each a raw model of its own unique presence and by its own means. The exhibition adapts to the environment in which it is produced. With the cloud database serving as a structural platform for the project,

each material extension of the exhibition becomes a prototype in and of itself, and as such, a tangible experience.

The project is a collaboration that takes place between two geographical locations, using the cloud as a communication channel, tool and archival form for the exhibition. New York and Berlin are both centers of global cultural production and significant platforms for local artistic experimentation. The participants come from a spectrum of disciplines including Design and Technology, Communication Design, Interactive Design, Fine Arts, Photography and Illustration. In June 2012, thirteen Parsons students travelled to Berlin to participate in a weeklong workshop at the UdK Berlin. In November of 2012, a group of students from the Digital Media class at UdK Berlin travelled to New York to complete the project and install the first of several iterations. A second version of the exhibition will be downloaded from the cloud and take place in Berlin in January 2013. The content of the show will remain dynamic as the participants continue to upload new versions of their prototypes. Future iterations can then be downloaded at additional sites globally.[1]

The resulting projects were highlighting the challenges and opportunities in creating digitally defined, but physically constructed artworks. In this case, the aim was indeed to create artworks, rather than product series so that the range of works only partially met the core of the "Beyond Prototyping" vision.

The three resulting works are summarised below and serve to illustrate the project.

Attracting countries is a static/localized data-sculpture dealing with the topic of migration (Fig. 1) . The centre represents the country where the exhibition takes place. It points out how it influences adjacent or more distant countries. The idea is to demonstrate how many people immigrate to the centre country, and where they immigrate from.

The construction contains 100 threads attached to a Plexiglas ring. The end of each thread is connected to a needle that floats horizontally toward the centre of the ring. They represent the relative count of immigrants and are attracted by a magnet, which is the exhibition country. Each needle has a small but noticeable distance from the magnet, which represents the path the travellers have gone to get to their destination.

Stamp lamp connects emotionally to its owner by using his/her biometric signature in the design (Fig. 2).

Fingerprints are used as an input to modulate the design. In this manner, the object is inherently connected to its "creator".

Most fingerprint identification systems do not look at the pattern of a fingerprint, but more commonly use certain points on the fingerprint for identification. These points are called minutiae, and their position to each other makes them unique. One kind of minutiae is called the bifurcations, meaning that one ridge on a fingerprint is divided into two ridges. These are the points that are in this case used to generate the form of the object.

---

[1]http://cloud.parsons.edu.

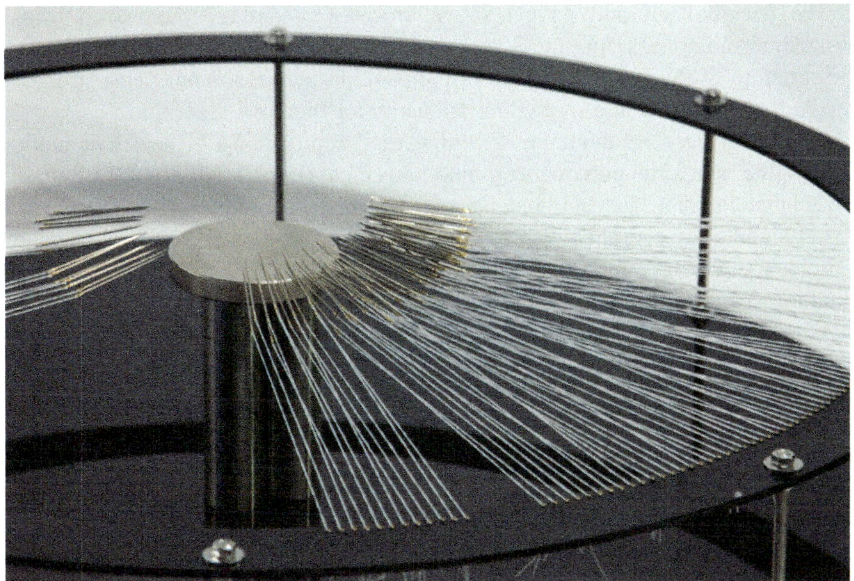

Fig. 1 ATTRACTING COUNTRIES by Felix Worseck

Fig. 2 STAMP LAMP by Gaspar Battha

The design itself and its generative process are strongly connected to the bifurcation minutiae. Using these "dividing" points according to their coordinates, the structure of the lamp is divided at the position of these points. This structural design is then pulled into three dimensions by a generative algorithm.

Signature piece is about the recording and reproducing of sensitive motion through the transformation of energy into movement (Fig. 3). A significantly simple device that can reproduce any one person's scripted signature on site. Utilizing the Cloud signature piece is able to send its DNA anywhere it so chooses, encoding individual signatures into its design as it is produced. Using the DNA, individual exhibitors can produce the machine on site via any rapid prototyping method, thus making the availability of this machine limitless and exponential.

All three pieces described above were created as a hybrid between material knowledge of the fabrication (using mainly laser cutting as the physical tool). All of them required a considerable effort in the physical assembly, even if the algorithmic design process, once established, was easily adapted to different contexts.

This was a concrete experience about how the seeming lightness of digitality, transmitted through the "cloud" was confronted with the effort and expertise required in actually assembling and installing the individual works. The whole course was designed to bring the participating students together in Berlin as well as in New York—not only in the mechanical sense, but also in a physical one as well. Having this physical connection, and really knowing the people you are working with on a project proves that a mediated collaboration can be both effective and rewarding.

Fig. 3 SIGNATURE PIECE by Andreas Picker

The experience also highlighted the challenge of finding meaningful designs and striking aesthetics with a meaningful "configurability", i.e. how to convert aspects of the design to parametric digital model where the form stays aesthetic with changeable narrative.

Ready/Made The next teaching offering between the UdK Berlin and TU Berlin focused on the threshold between mass manufactured goods with customisable parts through digital fabrication.

> The "Ready/Made" course looked at how rapid Prototyping Methods become more and more important in manufacturing products or product parts. Mass produced objects of high material variety and quality but with no individualised features coexist with highly customizable parts and objects limited by choice of material, size and surface quality. During the course students developed symbiotic relationships and computational models between a mass-produced and a customized object to create new ready/made hybrids that benefit from the best of both worlds using 3D printers, laser cutters and milling machines.

Since the course was offered as a part of the syllabi of the TU Berlin and the UdK Berlin, the challenge was to set course aims so as to be relevant to both career paths. Where the computer graphics department would expect developing ideas and solving the technical challenges within the realm of 3D geometry processing, either in capture or production, the UdK Berlin students' would be challenged with developing believable concepts, compelling aesthetics and meaningful narrative, along with the technical solutions.

Invase is a platform that enables the conversion of normal drinking glasses to flower vases, where the user can easily configure the type of flower arrangement and match that with the base glass form (Fig. 4) . An algorithm then generates a 3D geometry that can be 3D printed and attached to the glass.

Screw lock presents a system for "physical password security" wherein a screw and matching screw key are generated from a passphrase (Fig. 5) . Hence, the product affixed with the screw lock can only be opened by the holder of the key, or broken in the process.

With *HeroMe* system one could replace the head of a toy action figure with a 3D scanned head, enabling one to transform the figurines to personal doubles (Fig. 6) . To facilitate this, a Microsoft Kinect generated point cloud was semi-automatically converted to an appendix that could be mounted on the otherwise mass-produced toy.

Techno Legacy The course "Techno Legacy" took a look back in time at the history of innovation, challenging students to closely study a particular historical innovation and convert it to a relevant enquiry in today's digital context (Fig. 7). *"Mechanical Pi—In Memory of William Shanks"* is a machine that would mechanically operate an old calculator, using an algorithm that iteratively approaches to the value of pi in increasing number of decimal places.

Fig. 4 INVASE by Alyssa Trawkina and Marjam Fels

Fig. 5 SCREW LOCK by Gaspar Battha and Daniel Dalfovo

Kepler's Dream by Michael Burk and Ann-Katrin Krenz used 3D printing to create a fantastical abstract landscape inspired by the mystical world view of Johannes Kepler, made visible through an optical apparatus and mechanical gimbal, conjuring images from a medieval orrery (Fig. 8).

Fig. 6 HEROME by Nizar Ben Sassi and Robin Henniges

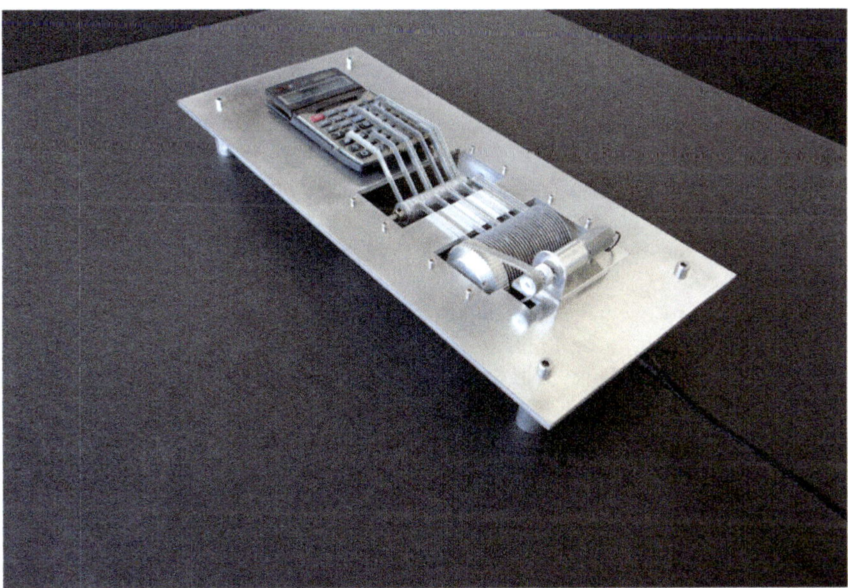

Fig. 7 MECHANICAL PI—IN MEMORY OF WILLIAM SHANKS by Florian Born and David Fröhlich

Fig. 8 KEPLER'S DREAM by Michael Burk and Ann-Katrin Krenz

Narrative Material Finally, the course "Narrative Material" explored a hard-coded meaning making into artefacts, as a counter reaction to the omnipresent screens in the public space, where the standardised form factor of a mass product dictates the aesthetics of the space. By creating spaces and objects that have the meaning physically encoded in them challenged the students to think hard about the message and the carrier to be seamlessly telling the same story.

Stephan Sunder-Plassmann created a park bench to remind the audience of the Nazi book burnings: A campaign conducted by the German Student Union to ceremonially burn books, in both Nazi Germany and Austria, by classical, liberal, anarchist, socialist, pacifist, communist, Jewish and other authors whose writings were viewed as subversive or whose ideologies undermined the National Socialist administration ideals (Fig. 9). "*Ort des Geschehens*" (The Place Where it Happened) recalls these crimes by linking the burnt books and persecuted authors at the actual place of event with a subtle memorial. A regular bench intended to be used as a place to sit, meet, think and read is slightly modified by implementing a barcode at the back. This code is the ISBN code of one specific publication, decodable using any conventional device, enabling the user to access any platform, which makes this book "readable" once again, and one is invited to sit down, read and reflect on the book at the actual place where it was burnt decades before.

These teaching projects served to guide and inform the "Beyond Prototyping" research efforts throughout the duration of the project, highlighting the technical challenges, aesthetic opportunities and simply inspirational case studies as to where to potentially apply digital fabrication as a part of a meaningful building process.

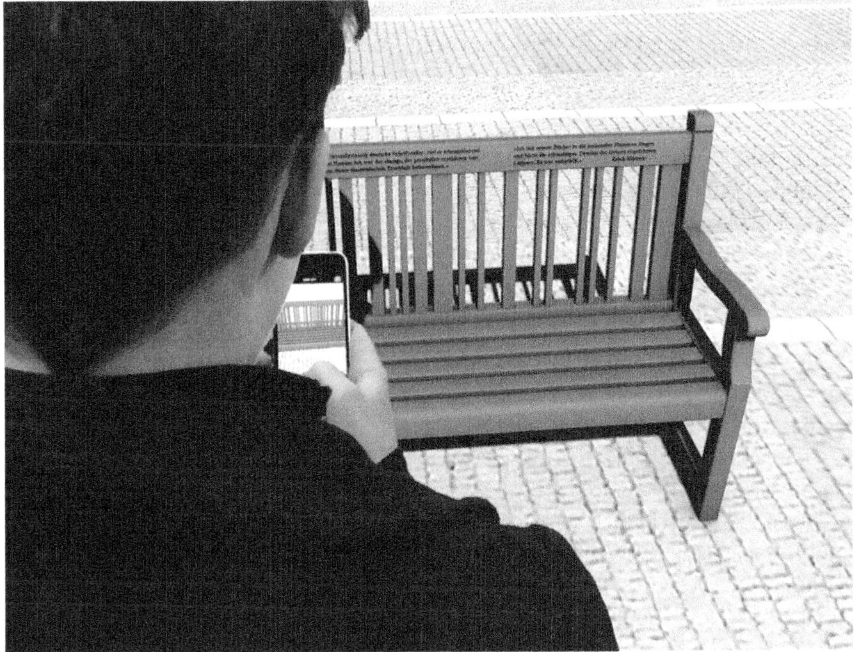

Fig. 9 ORT DES GESCHEHENS by Stephan Sunder-Plassmann

3 Case Studies

One of the motivations for the "Beyond Prototyping" research endeavour was the observation that the word "design" is often misused in supposedly interdisciplinary academic computer science communities such as Siggraph, TEI, CHI or UIST. Too often reduced to denote styling or used to justify otherwise unrelated fun, design is seen very differently depending on the perspective. This is certainly symmetrical in that any other discipline's concisely defined terminology is surely abused outside the field. However, this "lack of seriousness" about design was one contributing factor to our motivation in employing design more centrally in the process, not only as a subject of enquiry, but also as the means for it.

So, where "Beyond Prototyping" attempts to take a closer look at the opportunities digital fabrication can provide in defining new kinds of products (through services) that enable users to have a considerable influence in the physical appearance and function of the product through an algorithmic translation of the design concept, we decided that the best way to research this would be to try to implement real, functional services that anyone could potentially use. The design concepts would not only have to be believable but also to actually exist as services, because the biggest advantage of digital fabrication is the capability of producing every single product as a unique piece. Hence, it would not be enough to speculate on the user's aspirations about a particular customised product, but instead, the products would be made so that one could truly reflect their impact and potential.

We developed a wide range of design concepts, focussing on different aspects of the fabrication:

- A simple production line, where the number of actors involved in the production would be to kept to a minimum.
- A network of Berlin based small studios and workshops that would together produce a high quality product as a service, choreographed by an online service.
- A service model, where a custom hardware-measuring instrument would be used to scan a space and create a fitting design for it.

These different approaches required a highly transdisciplinary set of skills beyond the research project staff, so we collaborated closely from the very beginning with a number of experts in cabinet making, 3D milling, online service development and materials to guide our designs to be as realistic and as high in quality as possible.

This dialogue, which took place in various workshop visits, informal discussions and process tests, were paramount to designing not only a concept, or a prototype, but a realistic design, with a functional service to realise it.

The three design concepts, explained below in detail were the following:

- *Locatable*: a table top with a street map grid engraved on its surface, chosen by the user through an intuitive web service from the global openstreetmap[2] database, and produced through three Berlin based small studios.
- *Ciphering*: a 3D printed precious metal ring that encodes four digits on the surface as selected by the customer on its website.
- *Highlight*: a generative lampshade that is made with the help of custom lamp socket mount 3D scanner and an accompanying web service that enables an intuitive design of desired light distribution.

3.1 Locatable—Balancing Between the Craft and the Computational

Beyond process optimisation, digitally fabricated products promise personalisation and customisation in form, function or meaning. Designing such objects, the entire system needs to be considered as a whole: The interface and the parameters the users may access and the manufacturing process between machines and manual workshops constitute a network that enables quality in design and in realisation, which supersedes the possibilities of each step alone. Designing in the real world with small workshops as a methodology, a case study of a mass customised table is

[2]https://www.openstreetmap.org.

presented, "Locatable", with strong aesthetics, meaning and function, illustrating the potential of combining digital fabrication with traditional crafts (Fig. 10).

Introduction Digital fabrication is the focus of a lot of attention presently. The proliferation of 3D printing technologies through ever more affordable printers, as well as flexible and efficient online services is hugely influential in forming how the new generation of designers are thinking about design. 3D printing, which has first been profiled as prototyping, has maintained the focus mainly in the form over other material qualities. This has focused the development in ever-increasing abstraction between the digital and the material, where the printers can print more complex shapes, defined by the digital models, where the users have to think less and less about how the actual manufacturing takes place, how the machines actually work and if the manufactured object works in the real world. Yet, given this new accessibility, the 3D prints are increasingly expected to become the final products, as described by Hague et al. (2003), and Gershenfield (2008, pp. 3, 42, 79).

However, much of the quality of our products stem from their materiality: haptic feel, texture, weight, temperature and even resonant sound, all being qualities that are a key factor in the quality of a product of traditional handicraft and manual manufacturing, not only in terms of useful knowledge, but also as a motivation to design and create a product as discussed by Sennett (2008, pp. 163, 196–198). These qualities are often lost in the digital abstraction (ibid., pp. 59–65) and we often accept these shortcomings without question, as we have become accustomed to what is possible with these techniques. Many digital fabrication processes can

Fig. 10 Detailed view of a finished LOCATABLE depicting Schöneberg area in Berlin

manipulate a broader range of raw material than additive manufacturing methods: Laser cutting and 3D milling lets one process a broad range of material, but requires much more from the workshop and maintenance of the machines and does not fit well with the vision of the desktop 3D printer at home. These processes are already used quite commonly in small series or even mass production, but their intrinsic customisability is not so commonly applied. (For example, every individual item can be different without accruing any additional costs in adjusting the manufacturing process: one only needs to change the tool path).

This research presents a way of combining the digital malleability and the deep knowledge of the material. It defines a way of designing and producing high quality customised products that leverage the traditional handicrafts skills together with the digital algorithmic manipulation. The design process is a form of a dialogue in defining the meaning, the narrative, the aesthetics, the interface, the manufacturing process and choice of materials within the constraints of the meaningful, controllable, affordable and manufacturable.

Prior Work Pioneers in this field, such as *Nervous System,*[3] *Unto This Last*[4] or *Fluid Forms*[5] have been developing products for rapid manufacturing for several years. Nervous System design studio has been exploring the formal possibilities of rapid manufacturing in designing with biological forms, generated by algorithms and converted to jewellery. Albeit some of their designs are user customisable, their work is mainly designed by the designers using software, and the final forms are then fixed as final designs that the customers can buy in different sizes. Their striking aesthetics is only possible through the new manufacturing techniques, and their biological narrative is a meaning that works beautifully.

Fluid Forms designs focus on the personalising aspect of production on demand, such as 3D printed rings, where the inside hides a relief of a finger print, a 3D milled fruit bowl with an exaggerated elevation map of an area of the user's choosing, or a clock face that is laser cut in the shape of a city map—all instances which leverage the user's projection of meaning onto the objects. This personal narrative defines the object more than any functional aspect of the design.

Unto This Last[6] creates customised furniture, and performs the making process as a robotic ballet at the studio's workshop and showroom hybrid. Using 3D milling, *Unto This Last* creates functional objects that have been optimised for the manufacturing process with striking aesthetics. The different modular components, such as joints are easily combined, dimensions adjusted, and produced on demand, fitted to the customer's spatial constraints. Due to the 3D milling as a process, the choice of material is much broader than when relying on 3D printing, and the physical dimensions are much larger, making the creation of functional furniture possible.

---

[3]Nervous System, http://n-e-r-v-o-u-s.com/ Accessed 20 April 2015.

[4]Unto This Last, http://www.untothislast.co.uk/ Accessed 20 April 2015.

[5]Fluid Forms http://www.fluid-forms.com/Accessed 20 April 2015.

[6]http://www.untothislast.co.uk/.

Vision As the above examples illustrate, the manufacturing method is a central part of the aesthetic potential of any artefact. Designing the form finding and manufacturing process are both essential parts of the design. Instead of trying to constrain the work to a rapid manufacturing process that one might be able to print at home, there is a great potential in designing a network of services, be it through software, robotic manufacturing or even a handicrafts workshop, working together as an ecosystem, partially applying the principle of peer production to the production of physical products, as proposed by Reichwald and Piller (2009, p. 72), but with a greater emphasis on distributed manufacturing resources. This type of system-level thinking enables a balance to be struck between material qualities, dimensions, personalisation and the infusion of meaning, all (hopefully) within the constraints of the affordable and the manufacturable.

Seeing system design as an inseparable part of the work, the real world is used as a research method. Instead of using a good number of workshops and machines in different departments of our university, the design is done with small workshops and studios in the city that subsequently define our production network and implicitly operate within the constraints of a real world context.

A Case Study: Locatable Locatable is a dining table with a street network carved and filled with resin on it. It is meant to be ordered online with a simple interface, where one only has to specify a single address. The address defines the cutout area from the OpenStreetMap database, which is sent to a 3D milling studio. The tool bit has been previously defined for optimal aesthetic, and the vacuum table under the milling machine makes it extremely easy and fast to carve the street network on a wooden board. Once ready, the board is transported to another woodworking studio close-by, where the grooves are filled with epoxy resin and subsequently sanded for the final aesthetic, before sent to the client.

A map as a motif is useful and aesthetically appealing at the same time, a location is often emotionally charged, and both present an optimal content for personalised product. The idea stemmed from the observation from everyday life, where daily events and near future plans are often discussed in front of a wall mount map. Most of the time, these discussions do not include specific street address searches due to the familiarity of the place, but instead, the map reference is giving broad visual backdrop that illustrates distances and puts the plans in context. Hence, an abstracted, but relevant map of an area where one is located was deemed as desirable for creating a sense of attachment to the product as described by Auclair et al. (2005), and an ideal context was thought to be informal breakfast or dinner discussions at home. A table top surface fits this scenario very well, but had a central requirement of good haptic and aesthetic quality.

The interface through which one can define the final table online was kept very restrictive on purpose: The entering of the address is the only interaction possibility, resulting in an abstract preview of the cutout. The reason for this was to control the aesthetic, as defined by the designer. It was designed to let the user enter something personal and meaningful into the object without the possibility of inelegant and unforeseen outcome resulting in either a fall-off in user experience as mentioned by

Auclair et al. (2005) or complications in the production process as discussed by Willis et al. (2011). A limited number of parameters also promised better means to optimize the overall production workflow (Fig. 11).

To prototype this service, all aspects of design were worked on iteratively, in close collaboration with a team of experts: An online interface was designed to keep the interaction flow as smooth as possible, and provide the data needed for the milling in an easy to use format for the workshop, and to estimate the total length of the routing needed to keep the production speed realistic. In the milling workshop different milling bits, routing speeds, carve depths and vector terminations were explored to find the optimal aesthetic that is ideal for many different layouts of streets and still able to be produced within reasonable machining time.

In a similar fashion, different resins and inks, as well as the manual sanding process were tested and reviewed with the experts to come up with a high quality handcrafted aesthetic that still kept the man-hours to the minimum, and hence in turn make the production of the whole table a realistic proposal.

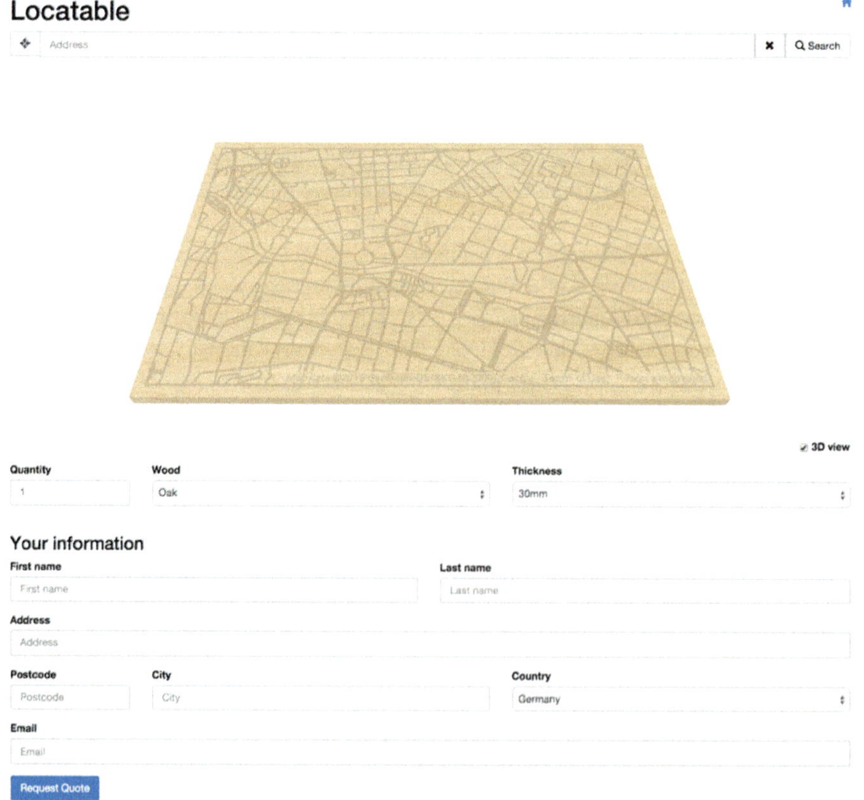

Fig. 11 Screenshot of the LOCATABLE order interface

A central question to the design process was the feasibility of the idea. This included the manufacturing and material costs, and the perceived value added from the end-user's point of view. To this end, it was decided to firstly push the concept to the extreme using an icon of cheap mass manufacturing as the base for our design, the Björkudden table from Ikea. We then carved our customisation layer on top, creating a dissonance between the mass manufactured object and the personalised service layer (Fig. 12).

The table was displayed to the general public as a part of the Berlin "Lange Nacht der Wissenschaften" (Long Night of the Sciences) event in the city, and a simple survey was conducted, asking the visitors to evaluate the table. The study was conducted only to give us an indication of the perceived value, and due to the nature of the event, we could not attract people to fill in complex forms, so the questionnaire was reduced to 3 simple questions:

1. What would you pay for the base product without engraving?
2. What would you pay for a generic engraving that you cannot choose?
3. What would you pay for a table with customised engraving that you can define?

The answers (N = 29) ranged significantly in evaluating the base product price, but overall, the difference between the base and the personalised was almost 100 %, albeit it did vary substantially (271 % sd = 191 %). In other words, our informal study suggests that people were willing to pay double for a table with a street map

Fig. 12 Close-up of the first finished table

carved and cast on it than the same table without the customisation. This indicative evidence would make the customisation feasible with our process for the mid price range upwards.

What many visitors noted was that once the base table was identified as an Ikea table, the value of the customisation was starkly reduced. The aura of the product was always seen as the combined value of the base product together with the personalisation, and the attached narrative. This experience led to the decision that using a mass manufactured base product makes little sense, as the meaning associated with the product conflicts with that of digital manufacturing.

Reflection The prototype in this process was on one hand the 3D milled, resin cast and manually sanded Ikea table showing the coordinates of the research group's offices, but on the other hand a manufacturing system that consists of a web component where one can define the design and order the table, the material supplier (and ordering process), the 3D milling studio and the woodworking studio for the different steps in the manufacturing.

Designing together with the experts in the studios proved an immeasurably positive benefit for coming up with the high quality aesthetics as well as providing direct indication of the feasibility of the process, how it would integrate in the everyday of the workshops and how much time it would take to actually produce the table.

Confronting the general public with the final result very much involved the pleasing quality of the table, where a transparent resin cast creates subtle ambivalence between the shadows of the grooves and the smooth haptic feeling. This reminds one of the importance of manifesting ideas in a physical space where these material nuances are extremely important in the final experience. It is easy to forget them in the complexity of software and process optimisation, as well as when concentrating on the narrative and the meaning from a more rational perspective.

Further Work Locatable was the first concept that was played through in the real world, and had tangible feedback from the everyday audience for the work. In parallel, various additional concepts are being developed that attempt to strike a balance between the narrative power for the individual, the manufacturing process mixing digital fabrication and manual processes and the definition of the system enabling the interactions. However, the feasibility of this approach is still highly specific and limited to a narrow range of products. As the pioneers in the field, *Fluid Forms*, *Unto This Last* and *Nervous System* all demonstrate, finding the meaning beyond the purely aesthetic is still a very difficult balance to strike. *Fluid Form's* focus on the personal story, at the expense of functionality, *Unto This Last's* focus on the spectacle of manufacturing, and use of 3D milled aesthetics, without additional narrative, and *Nervous System's* complex forms as aesthetics, but with very few meaningful personalisations all highlight the challenges of finding the right balance.

This text demonstrates the feasibility of a service and a network of production steps that may consist of the new or the traditional that together make possible aesthetics that cannot be yielded with digital fabrication alone. The network can serve as a design tool in the prototyping phase and be used for production as well. In case of increasing demand, and given a well-designed and well-documented workflow, the network can be extended easily to provide more production capacity by adding nodes, namely workshops and digital manufacturing studios.

The design process is far beyond simply form-finding, and is a balancing act between which aspects to limit and those that it makes sense to leave open to end-user manipulation. With the right mix, it is possible to find striking customisable products that can provide functional added value to the end-users and that are also feasible in smaller-scale production, and still beyond the safe haven of a university research project: designing in the real world.

3.2 Ciphering—Sense of Ownership of Generative 3D Printed Artefacts

This text looks at the infusion of meaning in digital fabrication processes. We analyse how non-expert consumers identify themselves with digitally manufactured products and whether the embedding of personal content can change their perception of a particular product. We present a case study of a customisable, digitally fabricated ring—*Ciphering*—which encodes personal information into a physical object. Through a website, the user can enter four digits (such as a date, for example), which are then encoded in the physical shape of the ring and only legible when held in front of a light source in a particular way.

In order to analyse the sense of ownership of this product, as well as to understand its appeal, we conducted interviews and surveys with customers. Instead of paying subjects to take part in the study—as is commonly the case in academic contexts—the design was implemented as a functional online service where people customised and purchased the ring. In this way, we could collect users' reflections in a real scenario, which was much more useful than speculating on an imaginary service. The study suggested that the narrative aspect, along with the sense of authorship, were central to the identification with this product. Additionally, we found that the meaningfulness of the parameters that customers can control, as well as the level of impact they have on the physical design, are both important aspects to take into consideration when designing a digitally manufactured product, and can allow the users to identify better with it.

Introduction Without question, the advance of additive manufacturing is changing the way we think about products: In manufacturing processes, it is possible to produce parts in quantities from one to several thousand, depending on the demand. In the design aspects, with no more tooling constraints, designers are free to create new shapes that were impossible to manufacture before (Campbell 2006). In fact, a growing number of companies are investing in systems using additive manufacturing (Wohlers and Caffrey 2013). Only in the past decade or so, this industry has developed notably from a mere prototyping tool to a real manufacturing system, able to produce complex high-performance outputs, such as aircraft parts (LaMonica 2013).

Along with the advancements in the area of industry, this technology is also creating dramatic changes in the way users perceive and relate to products. This can be observed in the proliferation of the so-called "maker" culture, as well as new formats of creating and sharing products, such as Fab Labs. These small-scale workshops equipped with computer controlled tools use digital fabrication to democratise manufacturing technologies previously available only for expensive mass production (Menichinelli 2011). Distributed manufacturing—a decentralised system using a network of geographically dispersed facilities—is also one of the phenomena developed hand in hand with information technology.

Other exciting aspects of this field are the newly available possibilities in the customisation of products. Customisation is the process of taking a general product design concept and tailoring it to the needs of a specific customer (Carter 2013). Customisation can be handled very easily in additive manufacturing in contrast to conventional manufacturing processes (Campbell 2006; Nambiar 2009). Since every product made with additive manufacturing can be unique, taking individual customers needs into account becomes possible while keeping mass production efficiency at the same time. This use of flexible computer-aided manufacturing systems for producing custom output is called mass customisation. The concept is attributed to Stan Davis in Future Perfect in 1987, but more recently, Tseng and Jiao proposed another definition. Tseng et al. (1996) define mass customisation as producing goods and services to meet individual customer's needs with near mass production efficiency. This possibility for the consumer to influence in the creation process of the product and become its own designer seems truly revolutionary (Carter 2013).

The advantages of mass customisation as a viable business strategy had been widely discussed (Piller 2005). For some time, the opportunities of mass customisation has been acknowledged as fundamentally positive by theoretical and empirical studies, and some companies are already having a degree of success employing this model. On the other hand, many companies have also failed in their attempts to implement it and large-scale mass customisation operations are still limited to a few examples (Harzer 2013).

One of the reasons for this, according to Piller (2005) is that most of today's offerings focus only on style, although in fact this option may be the least appealing to consumers. This problem is quite evident in many of the existing generative products using mass customisation strategies today. In areas like jewellery, for example, the customisation parameters that are accessible to the user seem often arbitrary; one can change the amplitude of a curve on a surface, radius of the bevel or choose a predefined selection of patterns. These parameters have little value for the user and hardly allow him or her to project a personal meaning into them. At the other extreme, many products provide the possibility of entering letters and create personalised cufflinks or pendants, or upload a picture to be embossed on the face of a ring, for example. In these cases, there is an attempt to introduce personal content, but being able to see the letters/images directly on the surface of the product is predictable; beyond the choice of the font or thickness of the lines, not much is actually designed.

Although mass customisation can increase the value of a product, this is not always necessarily the case. In fact, many mass customisation products do not create enough additional value for customers compared to their alternatives (Piller 2005; Squire et al. 2004). And as Campbell (2006) states, there is very little point in customising a product feature that will not add value. Therefore, one of the biggest challenges of the mass customisation field is to find meaningful parameters to customise, according to real needs and desires of users. We believe that design can play a vital role in recognising these motivations, creating products that are both meaningful and engaging.

Towards Meaningful Customization Every artefact we use already inherently bears a degree of cultural meaning embedded in its aesthetics and functionality (Siefkes 2012). Tables, for example, facilitate social gatherings and serve as a meeting point; rings often signify events, people, or affiliations. Therefore, the understanding of the cultural aspects of artefacts, as well as their functions and affordances, is fundamental for designing mass-customised products. Once the design goes along with these cultural and functional aspects, the customisation does not only make the process of buying the product easier, yet can actually enhance the experience of using it. This is exemplified in different prototypes designed as part of the "Beyond Prototyping" research project. One of the prototypes that serves as a study case for this paper is *Ciphering*, a digitally fabricated ring (Grimaldi et al. 2013).

Rings are interesting objects in terms of customisation because of their long tradition of hand-made production, as well as their small size, which makes them feasible to produce with digital manufacturing tools. The custom of giving and receiving rings dates back over 6000 years. As other types of jewellery, it has different cultural functions: it is used as a marker of personal or social status, as a signifier of some form of affiliation, or as a symbol of personal meaning. It often symbolises group membership or status. One of the aspects we found more interesting when designing a ring was the twofold affordance that this object possesses, allowing two different functions: On the inside, it carries an intimate piece of information—normally personalised in the form of a name or a date engraved inside—which can be seen only by the user. On the outside, visible to other people, the shape expresses its value in a symbolic form.

Ciphering: The Narrative Potential With this in mind, we have developed Ciphering, a digitally fabricated ring that encodes personal information into a physical object. In order to customise the ring, a user can enter four digits—a date, for example. In contrast to other generative jewellery, the digits are encoded in the physical shape of the ring and only legible when the ring is held in front of a light source in a particular way. In this manner, the ring keeps the two different functions: a personal message on the one side and a distinct aesthetic on the other (Fig. 13).

Since Ciphering provides users the possibility of designing a product according to a date—which represents a personal event—it most probably evokes emotions and personal memories in the user. This strong narrative aspect can play an

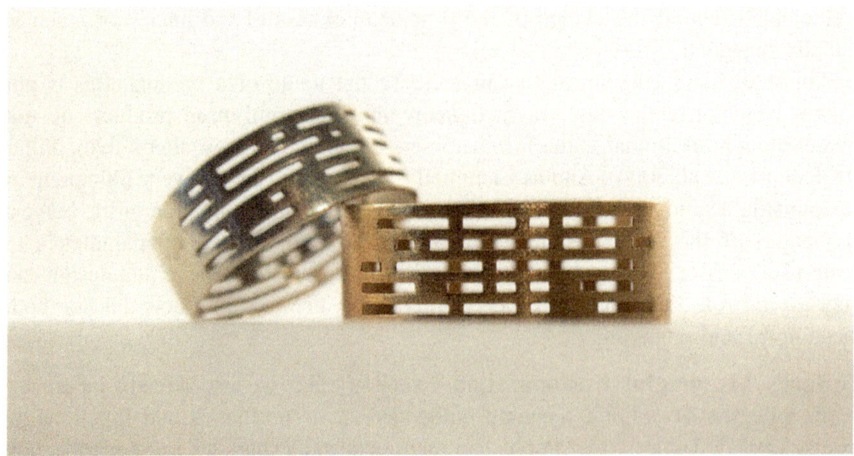

Fig. 13 A pair of finished CIPHERING rings in different materials

important role in how users identify with this product and contribute to their sense of ownership. In order to understand this phenomenon more in depth and recognise the narrative potential of Ciphering, we look at the theoretical framework proposed by Grimaldi et al. (2013). In the paper "Narratives in design", they analyse the different definitions of narrative as well as the roles and functions of narratives in products and design processes.

Grimaldi et al. also show how the narrative can contribute to the value of a product. This is exemplified by the project Significant Objects from Rob Walker and Joshua Glenn, which aimed to measure the added value that an accompanying story adds to an object (Glenn et al. 2012). In this project, they purchased cheap objects at flea markets and then ask writers to write an accompanying story for them. The objects were then sold on eBay with the attached story to verify the increase in value (for example a glass that was bought for $0.50 was subsequently sold for $50). As Grimaldi explains, the buyers were not purchasing the story, freely available online, but simply the object which acquired meaning through the story (Grimaldi et al. 2013).

We believe that one most important aspects that makes Ciphering valuable and appealing is in fact the narrative, as the customisation of the product is connected to a personal story. At the same time, we believe that the traditional narrative aspects of rings is enhanced, since Ciphering not only activates remembered or associated stories in the user, as rings normally do, but also includes the narrative as part of the design of the object.

Ciphering: The Service The customisation of Ciphering is done through a website (http://ciphering.me), which enables the ergonomic and material customisation of the product in a very easy and intuitive fashion. As the design concept is not obvious, a large portion of the content on the page explains the concept through 3D

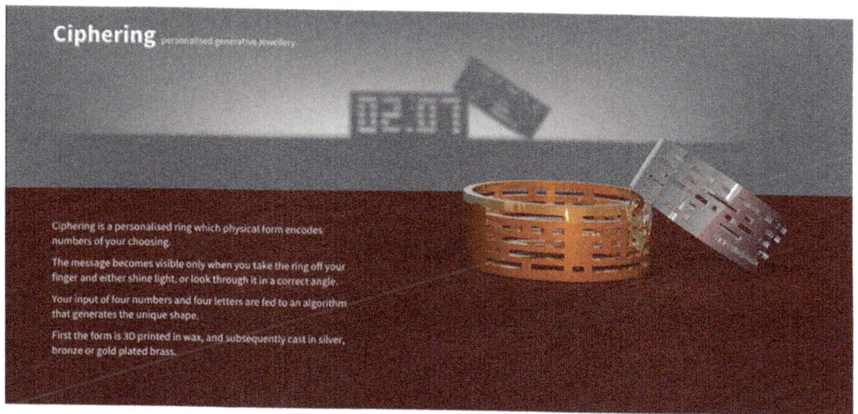

Fig. 14 CIPHERING website. http://ciphering.me/

renderings, photographic and video documentation as well as schematic illustrations. These are essential, especially since the customer cannot physically examine the ring, as would be the case in a traditional shop (Fig. 14).

All the information the customer can enter is done through simple text entry or choosing an option from a pull-down menu. These variables fundamentally affect the design of the resulting ring, but are exposed as precise choices for the customer. The entered parameters are piped to an OpenSCAD service that generates the solid 3D geometry ready for 3D printing. The four digits are converted to a five-pixel font, and scattered through the ring walls from a single vantage point. All the surrounding pixels are then randomly distributed either to the front or the back wall of the ring, disguising the digits into a pixel pattern. Only by looking through the ring from the projection vantage point, the pixel grids align, exposing the entered digits (Fig. 15).

The generated 3D model of the ring is automatically uploaded to a 3D printing service portal, from where the customer subsequently receives a confirmation email, when the geometry is processed and rendered for a realistic 3D preview. The customer can then see the final price (based on the choice of material and the total material volume of the ring) and order the product. The production and delivery, depending on the material, takes between 2 weeks and 1 month (Fig. 16).

A significant challenge of the service system was that the customer could not immediately try the ring on to gauge its ergonomic fit. This problem of uncertainty

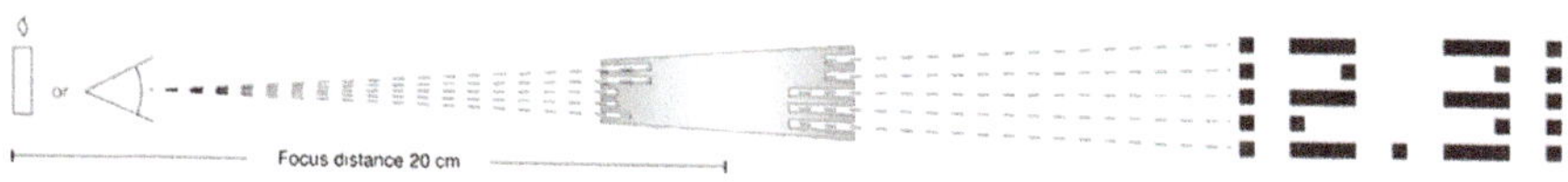

Fig. 15 How Ciphering works. http://ciphering.me/

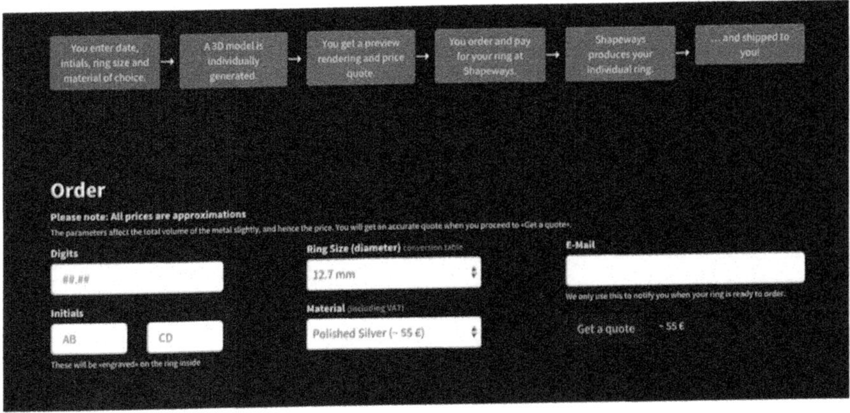

Fig. 16 Online configurator. http://ciphering.me/

was already devised in mass customisation systems and in many cases presents a disadvantage (Piller 2005). But for us it was interesting the way customers tried to overcome this problem: many went to a local jeweller just to measure the correct size. Alternatively, some have asked to have the ring produced in cheap plastic to confirm the fit before ordering the precious metal version. This of course delayed delivery of the product even further, which for many kinds of products would not be acceptable. However, since rings are acquired to represent an event planned well ahead of time or in retrospect, they are less problematic than other types of products.

User Survey In order to analyse the appeal and the meaningfulness of this product, we conducted interviews and surveys with the users. For this purpose, the design was implemented as a functional online service through which anyone could buy the product. As opposed to traditional user-studies, in which the subjects are paid to take part, we conducted this study with real customers. Individual users' motivations vary from person to person when they buying a product. We believe that user's reflections about their experiences in this real scenario carry much more weight than in a fictional scenario. Hence, making a real service available was the key to understand the true value of this product.

Designing the product, making it available and running the service were all parts of a system, and they would not make sense if they were not interconnected. By creating this fully working system in a concrete case, we were able to focus on the real experience, instead of posing questions or speculating. This process allowed us to understand which kind of aspects should be considered when designing digitally manufactured products.

The website went online in March 2014. The site was used (people filling their customisation details and asking for a quote) 367 times in 9 months. Thirty-two people bought a ring at an average price of $76 (prices ranged from $28 to $303, according to the materials used). Eleven customers filled the online questionnaire, which was sent by email. The questionnaire was structured in three parts: in the first part, participants had to enter general demographic information as well as answer a few key questions, such as whether they had experience buying jewellery online before, and what they found interesting about Ciphering, for example. The second part focused on their connection to the product and the meaning of the encoded message. The last part assessed their experience of using the ring.

Results Most of the participants (91 %), only half of whom has never bought jewellery online before, found the price fair or definitely affordable. 73 % of the users who answered the survey had bought the ring. The ones who did not buy it, reported that it was mostly because they were not planning to buy it in the first place or because they were not sure if it would fit their finger. But all agreed that the most appealing or interesting feature about Ciphering was the concept.

From the ones who bought the ring, all reported that the numbers had a special meaning for them and were related to personal events (such as graduation, engagement and marriage). 73 % of the participants reported that the ring was especially made for them and 45 % replied that they took part in the design of the ring by encoding the numbers. To the question of who they considered to be the designer/s of the ring (with a choice between: themselves, the designer, the programmer, or a computer) half responded themselves (in conjunction with another figure).

The different properties of the ring in order of importance across all respondents who bought the ring were the following:

- One can encode its own meaning
- Every ring is unique
- The numbers are "hidden" (not visible unless you know how to see them)
- The process is fun
- The materials are of good quality
- It is produced on demand
- The ring is created by a computer algorithm

Discussion Clearly the most important feature of this product was not the perception that it was designed by an algorithm, but rather the ability to participate in the design process by encoding a personal piece of information and seeing its significant impact on the physical form of the ring. The lack of interest in the algorithm and the focus on the uniqueness of the ring and the encoding of a personal story suggests that it is both the narrative aspect and the authorship of the user-as-designer that makes this customisation valuable. Of course, the automation of this process was enabled by the algorithmic design, but it was not perceived as *the* prominent characteristic.

With respect to the narrative aspects, all the participants said that the numbers had a special meaning for them and mentioned life events in order to explain why, confirming that the narrative plays an important role in the value of this product. Regarding the authorship, all participants described the experience of purchasing the product with words like "fun" or "exciting", which underscored the fact that they were involved in the process and that they feel positive about the experience.

Surprisingly, some participants responded that felt that they personally took part in the design of the ring (together with either the computer or the programmer). This reveals a new way that customers relate to products and a shifting role of the designer. Furthermore, many users contacted us to ask for additional customisation to match their individual needs.

For purposes of understanding these phenomena, we looked more in depth at these aspects through different theoretical frameworks of mass customisation: we analysed the experience the customers described through the sense of ownership, the customer co-design experience and the "I Designed It Myself" effect. Through these concepts, we explored how the level of involvement in making Ciphering could have enhanced the enjoyment of the process and the likelihood of bonding with it.

Sense of Authorship While the value of the outcome is important in mass customisation, many studies have highlighted the role of the experience itself in the perceived value of these products. Studies have shown that apart from the benefits that consumers get from mass-customised products, for example in reflecting their personal preferences, they also may derive benefits from the customisation process itself, that is, the activity of doing something by themselves is perceived by many consumers as self-rewarding and they experience joy during the co-designing task as a result of the fulfilment of a rewarding, artistic, and creative act (Mourlas and Germanakos 2009).

Merle et al. (2010) argue that from the consumer's point of view, the experience of co-design can have a positive influence on the overall value of mass customisation. They demonstrate that apart from an efficient customisation, there are complementary mechanisms that create perceived value in these products. Thus they identify two global components: (i) the product—with three dimensions: utilitarian value, uniqueness, and self-expressiveness—and (ii) the experience—with two dimensions: hedonic and creative achievement.

Piller (2005) also highlights the importance of the experience along with the outcome and explains that products that are co-designed may also provide symbolic (intrinsic and social) benefits for the customer. This co-design experience generates a sense of creativity and enjoyment in the user, in the accomplishment of a task. Another benefit of the co-design experience is the sense of ownership, which plays an important role in the evaluation of self-designed products (Turner et al. 2011). Mourlas (2009) describes the benefits of "pride of authorship". In this effect, the positive outcome of having created a satisfactory product on their own, instead of buying a standard off-the-shelf item, gives consumers positive feedback creating a feeling of pride. In this way, consumers would value the mass-customised product more than an identical off-the-shelf product.

Similarly, Franke et al. (2010) describe this phenomenon as the "I-design-it-myself" effect, when the value ascribed by individuals to a self-designed object incrementally stems from the fact that they feel they were the originators of the object. They explain that the economic value of self-designed products has often been attributed to two factors: preference fit achieved (which should be as high as possible) and design effort (which should be as low as possible). However, they suggest a third factor, which is "the awareness of being the creator of the product design". In their studies, they present evidence that the I-designed-it-myself effect creates economic value for the customer: Participants have the opportunity to design different products, enabling different degrees of design freedom and choices between self-designed items and standard ones. First, they demonstrate the I-designed-it-myself effect by showing that individuals are willing to pay more for a product when they are the originators of the design. Secondly, they confirm that the feeling of accomplishment acts as a mediator of this effect (Franke et al. 2010).[7]

In the context of the *Locatable* development as described above, we also found suggestive evidence for the same increase in the perceived value.

In summary, the frameworks described in this section can be useful for explaining the sense of authorship that users felt in making Ciphering, as well as their enjoyment in the process. They can also explain why, apart from the narrative aspect of the product, users felt a special connection with it. Although the customisation offered them the possibility to express certain level of creativity, we observed that users even felt comfortable enough to actually get in touch with us and ask for extra customisation. This phenomenon, in which users feel the need to customise over the default possibilities that were offered to them, not only confirms the sense of authorship they developed, but it might go beyond. We believe that these requests present an interesting case for mass customisation, as it demonstrates that, by being in the middle between mass production and an atelier service, customers can use the advantages of both.

Between Mass Production and Atelier Service In order to design an online configurator that is easy to use, many aspects of the design had to be pre-determined and unchangeable, whereas in a manual design process, they could have been easily changed. For example, the 3D printing resolution limits, as well as structural constraints in encoding the pattern onto the ring meant that a maximum of five (ascii) letters in five-pixel high font could be encoded on the surface. Defined by this limit, the design was then constrained to two digits separated by a dot, instead of a five-letter ASCII letter pattern. The interface thus seemed most suitable for encoding a date onto the ring.

However, several customers, after playing around with the online configurator, decided to contact us directly and request features not readily available through the website. For example, a group of graduating mathematic students from Swedish university wanted a memento of their studies, and thus asked if they could shift the dot in the configuration to encode the first three decimals of π onto the ring. Another customer couple asked to emboss the year on the outside, and engrave another date with year on the inside, the numbers symbolising the time they met and

[7]Shapeways Waveform Earrings. In: Shapeways.com. http://shpws.me/ClWa. Accessed 28 April 2015.

the wedding date. Several customers also wrote back suggesting expanding the design concept to other forms of jewellery. We found this very interesting, as it seems that these customers were using their own creative ideas to adapt the product to their needs, going beyond what the online platform offered. This direct customer exchange much resembles a consultation with an atelier service.

Commissioning an atelier to design an object, be it interior or jewellery, means that the client is hiring an expert to first understand the needs and desires of the client and then translate them to a satisfying experience. Purchasing a mass-produced product, on the other hand, means choosing between ready-made options that implicitly are also available to many other people. Thus, the broader aim of the "Beyond Prototyping" research project, to which Ciphering belongs, is to explore the opportunities that algorithmic design and digital fabrication can bring in between atelier service and Mass Production.

Ciphering provides a service that embodies the design language of a designer, but manifests itself in a bespoke unique instance through the interaction of the client on the web service, generating a unique design based on the meaningful parameters provided. Hence, the design work is done once, but is adaptive to each individual purchase. Since the data entry and the production are both automated to a great extent, the costs for such parametric designs are significantly below the prices of an atelier service and simultaneously accessible to a wider audience. Furthermore, special requests were easier to accommodate than in a one-off design process, since the production pipeline was already set up and the rest of the system could be readily used for the production. In this scenario, the configuration, which enabled affordable but personal products, encouraged some customers to consult the designer, perhaps creating an even stronger sense of ownership of the product through this dialogue.

Conclusion Digital fabrication is still a niche market. However, it offers new possibilities for both users and producers and can make a product more appealing and personal. In this text, we have argued that design can play a vital role in recognising users motivations and have analysed the appeal of these products through the concrete case study of a digitally manufactured ring. The data collected from our survey demonstrated that in addition to the uniqueness of the shape, the two most important aspects for users were: (i) the embedding of a personal story, and (ii) their involvement in the creation process, leading to a sense of authorship. This suggests that the physical shape (which 3D printers can easily create) and the production advantages that this technology offers are only a small part of the meaningfulness of these products.

Furthermore, we argued that the narrative aspects of objects should be taken into account when designing these products: by understanding their cultural connotation, as well as its narrative potential, it is possible to find meaningful parameters for customisation and enhance the experience of using them. We have also highlighted the importance of the process along with the outcome. We showed that the ability to influence the shape of a product by introducing personal content is perceived by the user as a self-rewarding activity and brings a sense of pride and fulfilment. The level of involvement in making Ciphering has potentially enhanced the enjoyment of creating the product and the likelihood of bonding with it.

Therefore, we claim that the user's ability to partially design this product by introducing personal and meaningful data has contributed to developing a sense of authorship and increased the value of the product. Furthermore, the desire of users to customise more than what was offered to them suggested that there are fertile grounds for exploring new opportunities in between atelier services and mass production.

Future Work In contrast to many other customisable jewellery pieces, Ciphering combines the literal meaning encoding with the aesthetic configuration. The fact that the literal meaning can be deciphered in the object was an essential part of the design concept. We have demonstrated that the engagement in the process creates a strong sense of ownership of the product, but further work is required to discover how important the deciphering act actually is for the user. One can argue that many of the benefits of Ciphering also apply to generative products that cannot be "deciphered" afterwards, be it encoding an audio recording[8] or GPS coordinates to polygon mesh.[9] Hence, a further study is needed to clarify how important the functional aspect of the customisation is in relation to the personal engagement.

Furthermore, since the concept is automated as production pipeline and the design is in the algorithm, we intend to study the monetary value of the immaterial, for example with the pay-as-you-want model, to find the difference between the manufacturing and the narrative, when clustered between the customised, pre-configured or hand-made.

We continue to explore further opportunities in between atelier services and mass production through concrete working study cases in the form of usable tools, which facilitate the dialogue between users and designers. Together with two additional prototypes, the research project "Beyond Prototyping" explores not only how the designer's aesthetics and the users' needs are brought together with the help of custom hardware and software, but also what kind of new service models are best suited for mediating this interaction.

3.3 *Highlight—A Generative Lampshade*

Highlight is a digital fabrication service that creates custom designed lampshades. Each lamp is customized to a specific space and allows the user to direct the light to particular areas in the room. This way, the service combines the uniqueness of an atelier solution with the advantages of a mass-production process.

[8]See Footnote 7.

[9]Meshu. In: Meshu. http://meshu.io. Accessed 28 April 2015.

Fig. 17 An instance of HIGHLIGHT installed in a room

Every space we live in is different: not only its architecture, but also the way we arrange it according to our needs. When it comes to light, one might want to have a spotlight at the couch table, illuminate a piece of art on the wall, or highlight a specific feature in the architecture of the room. With this in mind, we developed Highlight, a digital fabrication service that creates custom-designed lampshades, allowing users to direct the light to areas they feel important.

The era of digital fabrication brought new customization possibilities. Using algorithms, designers are able to generate products according to specific needs of users. In this line, designing objects that respond to every interior seems like an obvious step. However, this process has always remained challenging, as it requires designing new tools and systems as well as blending the physical with the digital. With this project, we address this challenge and demonstrate that with the assistance of technology, the designer's aesthetic can be adapted to a user's personal needs (Figs. 17 and 18).

Description The system consists of three core components: a custom built 3D scanner, a web-based generative software system and a 3D printing service. Initially, the user receives the 3D scanner as a loan. The scanner can be screwed directly into the existing lamp socket of the room and operated through the light switch. After the room is scanned, the users can see a 3D representation of the room in a web-based application and decide to which areas the light will be directed. From these data, the shape of the lamp is automatically generated and ready to be printed. Once installed in the room, the lampshade creates a special atmosphere, combining both diffused and direct lighting into one single object (Fig. 19).

Fig. 18 The directed illumination on the walls from HIGHLIGHT lamp

Fig. 19 A service description of the HIGHLIGHT system

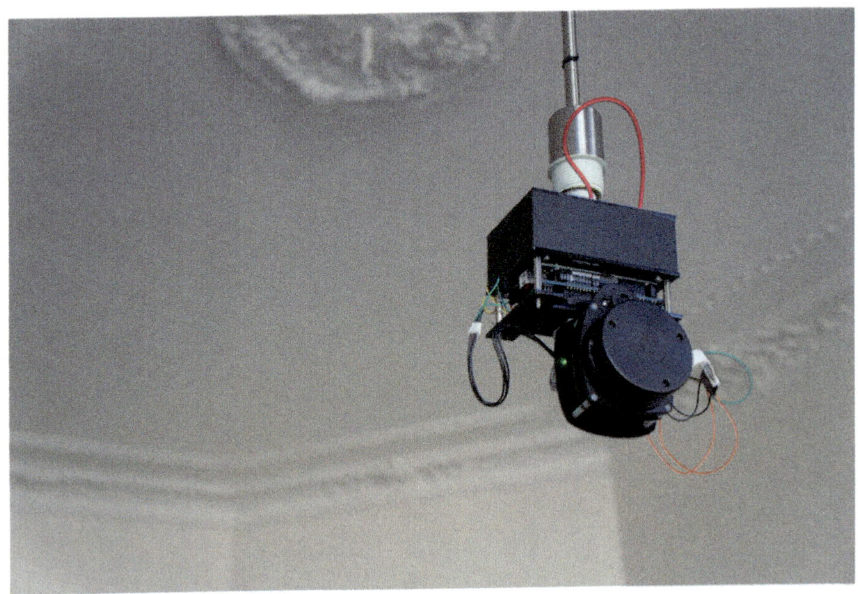

Fig. 20 Scanner installed to the lamp socket for 3D scanning of the room

Highlight takes advantage not only of the aesthetic and functional possibilities of digital fabrication, but also presents a new paradigm as a service, bringing together the uniqueness of an atelier solution with the advantages of a mass-production process. For example, since a 3D scanner is an expensive piece of hardware, it is provided to the users on a short time loan (Fig. 20).

Another big advantage of this service is the fact that a 3D preview of the lamp is generated in real time. The opportunity to see the lamp in the real context gives a better understanding of how the product will work than when buying it in a shop. This last point is of even greater importance when buying 3D printable products. In this case, one cannot even see the real object, but only a model, broadening the sense of proportion and scale. Hence, by connecting the model with the space for which it is designed, the product becomes more "tangible" (Fig. 21).

The intuitive browser interface is designed to allow users creativity and freedom. By using a "painting tool" as an interaction affordance, users can easily focus on the desired effect. The visual quality of the 3D point cloud both provides an implicit freedom to highlight the features of the space and at the same time also presents an easy-to-operate digital environment. The scanning of the room and its visualization in 3D serves as a design tool for users in a way for which they have never had access before. This offers them the ability to use their creativity and presents them with the opportunity to perceive their domestic spaces in a refreshing way.

The object has a strong narrative and performative aspect from a user's experiential point of view. In the process of "making", this object becomes meaningful and personal, and as a result, every lamp is strongly connected to the person who

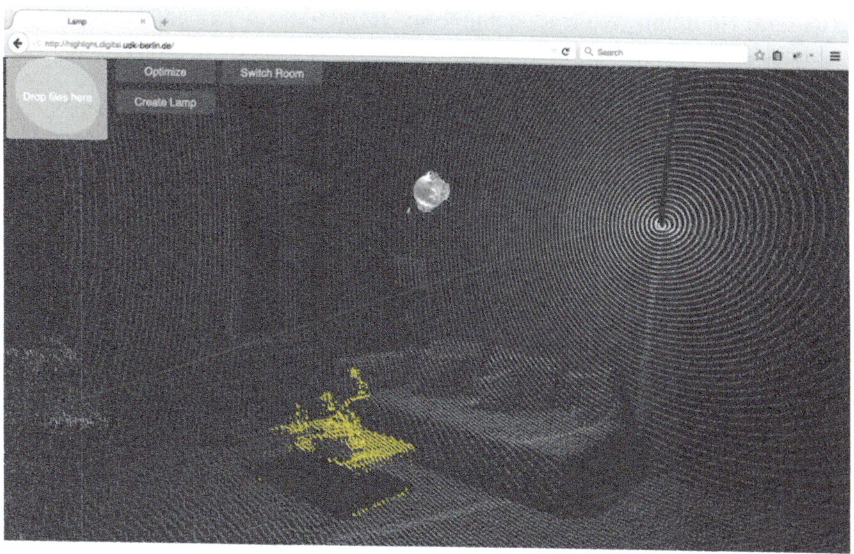

Fig. 21 The web interface for intuitively design the lighting based on the 3D scan

created it and the space it inhabits. The product is not simply a single object anymore, but goes well beyond this, including the whole process, from planning to realization.

In summary, the research project explores the possibilities of designing in an existing space in a tightly coupled way, where the designer's aesthetics and the users' needs are brought together with the assistance of custom hardware and software. The resulting striking and functional digitally fabricated artefact fits seamlessly in its context. The first iteration of the idea is a lampshade generator, but the concept easily translates to various other aspects of spatial design: from furniture to space dividers, wall surface materials to acoustic elements, and beyond.

Technical Details In order to scan the room in 3D from the location of a lamp socket, we developed a custom piece of hardware. Combining Robopeak's "RPLIDAR 360 Degree Laser Scanner" with a "Dynamixel Robot Servo", it is possible to conduct a volumetric scan of the room. A specially designed power converter provides power supply directly from the lamp socket, enabling a cable free installation of the scanner. The hardware is controlled by an Arduino based system custom software that saves the data on an SD card for easy handling.

The user can upload the data to a custom website, which converts the data into a point cloud. This serves as the basis of the user interaction, navigating a virtual representation of the room and selecting areas that should be illuminated. The software runs on the client side and is a custom solution built with THREE.js/WebGL/HTML5. This cross platform application generates the geometry of the lamp that can be viewed in real time and is exportable for 3D printing as an STL file.

Extending Highlight While scanning users' room offers benefits in terms of costs and ease of setup, it is potentially difficult for users to imagine the final result of the custom made lamp. Therefore, we extended the simulated approach by providing users with a physical lamp shape that can change the permeability of its sides. The individual sides can be manually controlled using custom software, offering users the possibility to create different light situations on demand.

We created two different prototypes of controllable lamps, with increasing geometric complexity to fulfil users' needs. The first lamp is created from a laser cut acrylic glass frame and 10 individually controllable sides from liquid crystal shutter panels. The second prototype consists of a laser cut base and 24 controllable faces, each being created from polymer dispersed liquid crystal (PDLC) switchable diffuser. The shutter panels and the switchable diffuser alter their transparency when voltage is applied (Fig. 22).

Description The lamps serve as a "live preview" for users on how different light situations will appear with their custom made lamps. The lamp is screwed into existing light sockets and can then be controlled using a smartphone or desktop application. Users can model their lighting environment, highlighting parts of their room or physical objects. The light setting can easily be maintained for later fabrication of non-dynamic lamp shapes. The physicality of light change provides users with a one-to-one mapping of their imagined lighting environment and the

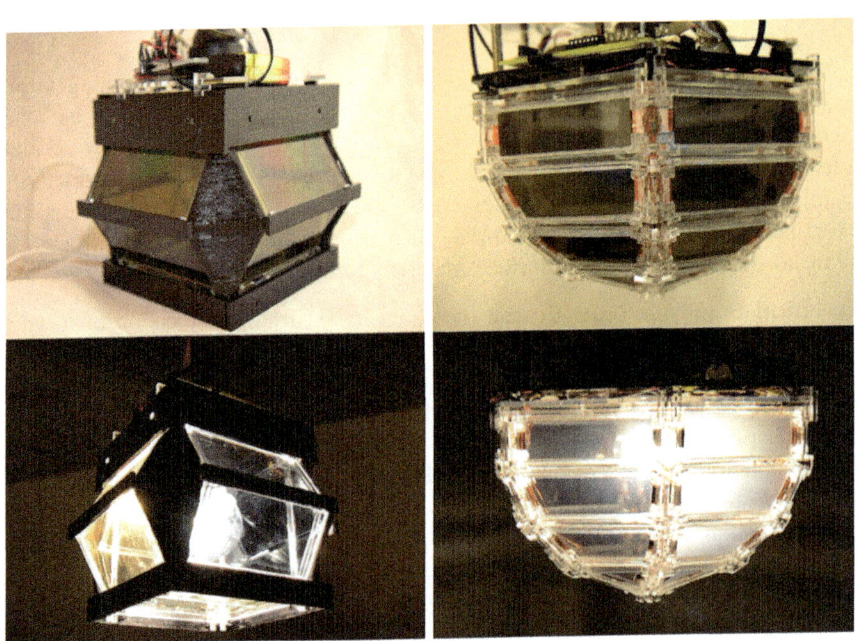

Fig. 22 Two different, complex geometric prototypes were created; a lamp with 10 LC shutter panels (*left*), and a lamp with 24 faces from PDLC switchable diffuser (*right*). Each face of the lamp can be controlled individually when voltage is applied (*bottom*)

later fabricated lamp. This eliminates the need for the creation of a virtual room as well as indirect manipulation of room light via the digital model. Manipulation is direct and users can also choose to explore light under different conditions of daylight or night.

While it would be possible to provide users with the dynamic version of the lamps, this is not necessary since users can use the device one time to decide the preferred lighting and manufacture a permanent lamp shade reproducing their favourite lighting environments. Additionally, this highly reduces costs. As with the simulation-based version of Highlight, users are provided with an expensive hardware for individualisation, which can later be exchanged for more cost-effective, manufactured versions (Fig. 23).

Technical Details The individual faces of both lamps are attached to a laser cut acrylic glass frame. The LC shutter panels are fixed in size and shape, therefore

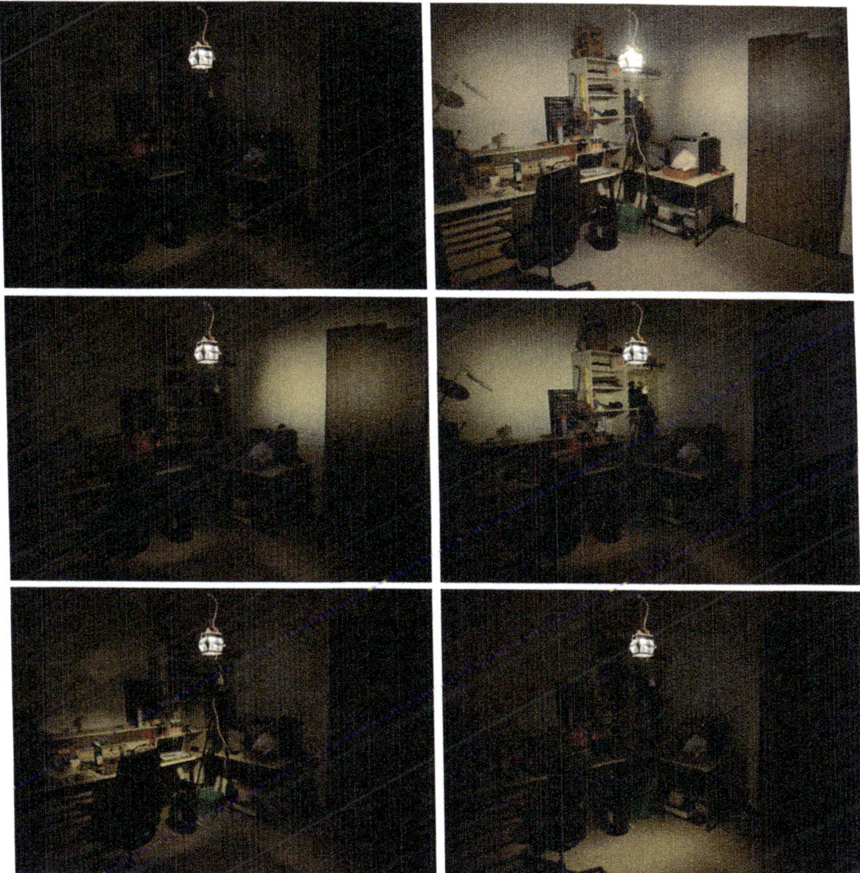

Fig. 23 Users can control the individual faces of the lamp to create desired lighting situations

Fig. 24 The LC shutter panels are mounted on an acrylic glass frame and connected to custom circuitry

allowing for creation of a limited variety in terms of lamp shape. The panels are actuated with 3–5 VDC and can alter their transparency from transparent (0 VDC, 50 % light transmittance) to completely opaque (5 VDC) continuously. This makes the panels especially suitable for controlling light permeability.

The second prototype consists of 24 faces of switchable diffuser, a material that turns transparent when voltage is applied (110 VAC). The material consists of two layers of conductive transparent material (ITO) with liquid crystals sandwiched in-between. This makes the material feasible to cut in arbitrary shapes using a laser cutter. We create 4 different shapes of switchable diffuser and enclose them in housing for protection and improved mounting. The custom circuitry controlling the voltage is created using a microcontroller, also including Bluetooth for remote control. Users literally connect their smartphone to the lamp to control it. The desired lighting environment could then potentially be saved and uploaded for manufacturing of the lamp (Fig. 24).

4 Beyond "Beyond Prototyping"

The great number of ideas and functional prototypes stemming from the courses taught during the research project, as well as the three case studies developed to functional services, present a vision for algorithmically defined products where the dialogue between the designer, the manufacturing process and the customer can be structured differently than before. Five years after the first discussions leading to the subsequent research proposal and the beginning of the project, today, many startups and more established players are developing ideas in a similar vein. 3D printing services are increasingly ubiquitous in creating small-batch products for the growing market, yet the algorithmically defined, customisable objects remain a tiny minority comparatively. Our research is very much aligned with this, having realised how difficult it is to find elegant design ideas where the customisation remains meaningful, not simply a manifestation of the latest technological trend or the mere capability of being able to do so, but something that is meaningful to the customer in more timeless manner.

On the other hand, many furniture startups are connecting local manufacturers as a network of actors for enabling local production in on-demand basis. In this field, also, the parametrically defined objects remain limited.

The case studies developed as a part of "Beyond Prototyping" are ready to be taken beyond the research and implemented as startups or integrated as designs with existing manufacturers. Meanwhile, the approach to use on-demand productions as a research tool whereby the audience are more invested in the designs has proven a useful tool for taking design more seriously in the academic context, not only as a subject of study, but as a proactive method of enquiry. To setup such working systems requires much more effort than prototypical one-offs, but provide more credibility in analysing prospective consumer responses.

The challenges in developing these systems we faced were considerable: On one hand, as with any software relying on public APIs of online services, we had to keep on updating the code so that it would stay compatible with the service providers. On the other hand, in integrating with small studios' daily work, the research project tended to have less of a priority than the client's orders, and hence finding time and space for the development was always a challenge. But these challenges were essential in creating a more realistic designs that would not only work "in principle", but also in practice, and through this practice, the prospective consumers, and hence our research subjects, could interact with the systems and share their insights with us.

In the end, out of the three concepts, only one, Ciphering is currently fully functional, as the Locatable network of participants still requires considerable human effort in coordinating everything, and in similar vein, the Highlight concept has the prerequisite of shipping of the scanner to the prospective clients, also demanding continuous human involvement. Ciphering, as it has been streamlined to directly interface with the 3D printing service Shapeways, can operate autonomously. In this case, even though there still is considerable human involvement, it takes place under the auspices of the Shapeways organisation, and does not involve any action from within the research team.

We feel that everyone involved, the students, faculty, research assistants, the small Berlin workshops and perhaps the wider design community as well have all learnt a lot through this project, and we remain grateful for this opportunity.

References

Auclair, I., Lalande, P., & Dorta, T. (2005). The influence of 3D modelling and rapid prototyping techniques on customized objects in industrial design. In *Nordic Design Research Conference, Copenhagen* (pp. 27–31).

Campbell, R. I. (2006). Customer input and customisation. In N. Hopkinson, R. Hague, & P. Dickens (Eds.), *Rapid manufacturing: An industrial revolution for the digital age* (pp. 19–36). West Sussex: Wiley.

Carter, A. C. (2013). *Mass-customization through digital manufacturing*. Master's Thesis, Auburn University, Auburn, Alabama.

Franke, N., Schreier, M., & Kaiser, U. (2010). The, "I designed it myself" effect in mass customization. *Management Science, 56*, 125–140.

Gershenfeld, N. (2008). *Fab: The coming revolution on your desktop—from personal computers to personal fabrication*. New York: Basic Books.

Glenn, J., Walker, R., & Grote, J. (2012). *Significant objects*. Seattle: Fantagraphics Books.

Grimaldi, S., Fokkinga, S., & Ocnarescu, I. (2013). Narratives in design: A study of the types, applications and functions of narratives. In *Design Practice. Proceedings of the 6th International Conference on Designing Pleasurable Products and Interfaces, Newcastle upon Tyne* (pp. 201–210). New York, USA: ACM.

Hague, R., Campbell, I., & Dickens, P. (2003). Implications on design of rapid manufacturing. *Proceedings of the Institution of Mechanical Engineers, Part C: Journal of Mechanical Engineering Science, 217*(1), 25–30.

Harzer, T. S. (2013). *Value creation through mass customization: An empirical analysis of the requisite strategic capabilities*. PhD Thesis, Aachen University, Aachen.

LaMonica, M. (2013). *Additive manufacturing*. USA: MITS Technology Review.

Menichinelli, M. (2011). *Business models for fab labs*. http://www.openp2pdesign.org/2011/fabbing/business-models-for-fab-labs/

Merle, A., Chandon, J.-L., Roux, E., & Alizon, F. (2010) Perceived value of the mass-customized product and mass customization experience for individual consumers. *Production and Operations Management, Volume 19*(5), 503–514.

Mourlas, C., & Germanakos, P. (2009). *Mass customization for personalized communication environments: Integrating human factors*. New York: Information Science Reference.

Nambiar, A. N. (2009). Mass customization: Where do we go from here? In *Proceedings of the World Congress on Engineering Vol I, London* (pp. 687–693).

Piller, F. T. (2005). Mass customization: Reflections on the state of the concept. *International Journal of Flexible Manufacturing Systems, 16*, 313–334.

Reichwald, R., & Piller, F. (2009). *Interaktive Wertschöpfung: Open Innovation Individualisierung und neue Formen der Arbeitsteilung*. Wiesbaden: GWV Fachverlage GmbH.

Sennett, R., & Michael, B. (2008). *Handwerk*. Berlin: Berlin Verlag (English original: Sennett, R. (2008). *The craftsman*. Yale University Press).

Siefkes, M. (2012). The semantics of artefacts: How we give meaning to the things we produce and use. *Image. Zeitschrift für interdisziplinäre Bildwissenschaft,16* (Special issue Semiotik). http://www.gib.uni-tuebingen.de/image/ausgaben?function=fnArticle&showArticle=218; Part 2.

Squire, B., Readman, J., Brown S., & Bessant, J. (2004). Mass customization: the key to customer value? *Production Planning & Control: The Management of Operations, 15*, 459–471

Tseng, M. M., Jiao, J., & Merchant, M. E. (1996). Design for mass customization. *CIRP Annals—Manufacturing Technology, 45*, 153–156.

Turner, F., Merle, A., & Diochon, P. F. (2011). How to assess and increase the value of a co-design experience: A synthesis of the extant literature. In *World Conference on Mass Customization, Personalization, and Co-Creation: Bridging Mass Customization and Open Innovation, San Francisco*. <hal-00649498>

Willis, K. D. D. et al. (2011). Interactive fabrication: New interfaces for digital fabrication. In *Proceedings of the Fifth International Conference on Tangible, Embedded, and Embodied Interaction, Funchal, Portugal* (pp. 69–72). New York: ACM.

Wohlers, T., & Caffrey, T. (2013). Additive manufacturing: Going mainstream. *Manufacturing Engineering*, 1–5.

The Results of Rethinking Prototyping

Jussi Ängeslevä, Benjamin Bähr, Boris Beckmann-Dobrev, Ulrike Eichmann, Konrad Exner, Christoph Gengnagel, Emilia Nagy and Rainer Stark

Abstract The scientists and academics in the transdisciplinary project called "Rethinking Prototyping" have not only been working on concrete hybrid prototyping approaches in their research, but also on a joint understanding and a general concept of prototyping as well. A differentiated analysis of the terms used in contexts connected with prototyping led to the finding that their application both differs from discipline to discipline and is partially complementary, too. In the transdisciplinary context of complex interrelated developments, it is not expedient to attempt a

J. Ängeslevä
Institute for Time Based Media, Berlin University of the Arts, Berlin, Germany
e-mail: jussi.angesleva@iki.fi

B. Bähr
Quality and Usability Lab, Technische Universität Berlin, Berlin, Germany
e-mail: baehr@cs.tu-berlin.de

B. Beckmann-Dobrev · K. Exner
Industrial Information Technology, Technische Universität Berlin, Berlin, Germany
e-mail: boris.l.beckmann-dobrev@tu-berlin.de

K. Exner
e-mail: konrad.exner@tu-berlin.de

U. Eichmann · E. Nagy (✉)
Hybrid Plattform, Berlin University of the Arts, Berlin, Germany
e-mail: emilia.nagy@hybrid-plattform.org

U. Eichmann
e-mail: ulrike.eichmann@hybrid-plattform.org

C. Gengnagel
Institute of Architecture and Urban Planning (IAS), Berlin University of the Arts, Berlin, Germany
e-mail: gengnagel@udk-berlin.de

R. Stark
Industrial Information Technology, Virtual Product Creation, Technische Universität Berlin, Berlin, Germany
e-mail: rainer.stark@tu-berlin.de

C. Gengnagel et al. (eds.), *Rethink! Prototyping*,
DOI 10.1007/978-3-319-24439-6_11

definition that will cover all prototyping concepts. Rather, the prototyping methods and concepts should be placed and described in a multi-dimensional matrix. This article discusses considerations in this regard and presents their reflection in the "layer cake" publication format.

1 Introduction

The hybrid prototyping approaches in the sub-projects of the main research project called "Rethinking Prototyping" arose from a particular multi-perspective constellation that has survived all-too-rarely in the long run: the engineering disciplines at the Technische Universität Berlin (TU Berlin) and the artistic-design disciplines at the Berlin University of the Arts (UdK Berlin) worked together to jointly develop new prototyping concepts from the very outset.

In the three-year project, scientists and academics addressed the latest prototyping concepts in order to perform an experiment that involved the creation of a general, transdisciplinary concept of prototyping. They reflected on their own methods and approaches, considered new ones and determined similarities and differences in the areas of use, complexity, materiality and the objectives of the distinct prototypes and prototyping processes.

The experimental aspect of the "Rethinking Prototyping" project was considered in the practice of collaboration in the sub-projects; all the participants engaged in a continuous exchange on a joint level in various formats for three years in the guise of colloquia, retreats, workshops and review conferences that provided opportunities for productive transdisciplinary exchanges (cf. Eichmann and Nagy in this volume). In the summary of the latest developments in each case, it was possible to find starting points that opened up theoretical discourse on the one hand, but also led to cooperation with practical results on the other. This cooperation, for example, enabled a research team from "Hybrid Prototyping" and "Blended Prototyping" to create a prototyping app for smartphones. The app lets users find suitable prototyping processes for their development tasks and to do so in accordance with their particular stage of development and the desired function of the prototype. The information for the design of the app's content was provided by the findings from the project-accompanying discourse on prototyping, which defined the intensive, three-year collaboration between the participants. How the search for an overarching definition of prototyping was designed and what results it brought are presented in the following.

2 Explanation of Terms: Model Versus Prototype, Design Prototype Versus Technological Prototype

Prototyping of technical and digital systems, products and design artefacts or components is one of the core disciplines in design and engineering. Nevertheless, major differences exist with respect to the motivation, use, function and goal of prototyping, as well as the degree of rigour in planning, executing and reflecting prototypes, which eventually represent the output of prototyping. In the project, it was possible to question the actual way prototyping is applied in the different disciplines.

This diversity of prototyping concepts was also reflected in the project by the diversity of terms that the representatives from complementary disciplines used. The research group recognised the need to define the terms at the outset in order to precisely describe the prototyping concepts. This was followed by the need for exchange based on concrete examples where the relevant characteristics for the differentiation of the terms are manifested. A portion of the discourse therefore shifted to practice and was reflected in the hybrid prototyping concepts of the partial projects.

In the theoretical discourse on prototyping, two complementary main forms of prototypes have been differentiated and described. They are called the *design prototype* and the *technological prototype*, a distinction that is generally made across disciplines. The differentiation of their content in individual fields is not identical, however, and it is not possible to clearly assign these two forms to specific disciplines. At the start of the project, the two forms were juxtaposed on the basis of the main functional areas of prototypes that were identified in the discussions and in the joint prototyping processes. These main functional areas of prototypes can be divided into four categories: (1) generating ideas and externalisation, (2) determining user perspective and expectations, (3) validation and testing and (4) communication. In these categories, one can identify numerous individual functions that are comprehensively outlined in the chapter entitled "Perspectives on Future Prototyping—Results from an Expert Discussion" in this volume and in the conference paper bearing the title of "A transdisciplinary perspective on prototyping" (Exner et al. 2015). Some examples of these functions of prototypes are:

A prototype

- visualises mental ideas;
- supports the comprehension of complexity;
- enables communication, thus removing cultural and linguistic barriers;
- always contains a specific question and is limited due to given constraints;
- tests functionalities and requirements;
- creates a basis for common understanding of the idea that should be realized;
- localises users' interests and/or
- allows analysing users' interaction with the object.

To begin with, the two main forms of prototypes can be described on the basis of the categories and their associated functions as follows:

Design Prototype At the beginning of a development process, a design prototype serves to externalise an idea, determine the target horizon and define the problem. In later development phases, a design prototype also primarily involves functionality, but the aspects of use, interaction and communication take precedence here. Questions about the acceptance and the needs of users as well as the complexity and sequence of actions should be answered on the basis of the prototype. Adjustments and the consideration of alternative design proposals can be easily included since a wide spectrum of design and layout options remain available in this stage of development. Lastly, design prototypes also answer aesthetic questions.

Technological/Functional Prototypes An essential aspect of technological and functional prototypes is to prove the functionality and the implementation of the planned and developed product. This usually occurs in the late development phases. Only a few options for alteration in the specifications remain at this stage of development, since the effort and costs of adjustments rise disproportionately. The main objectives of functional prototypes in today's engineering approach are to evaluate the results of the development process and to ensure preparation for serial production of a product. However, in tomorrow's engineering approach the interplay of interdisciplinary teams need new types of functional prototypes early on the engineering process. Currently, such new functional prototypes are under research and development.

To summarise, it is not possible to assign a specific prototype to only one discipline. In this context, it was essential to research the interrelation between the different types of prototyping within the involved disciplines. By addressing precisely these issues, "Rethinking Prototyping" started where these traditional dichotomies of the two complementary prototype concepts can no longer hold sway and their merging in hybrid processes is necessary. Today, for example, development tasks in design are solved generatively and individually, which can give rise in principle to an endless number of prototypes that may also simultaneously be understood as a product. At the interface between the algorithmically-generated design and traditional design, the sub-project called "Beyond Prototyping" pursued research related to quick production possibilities in the creation of individualised products. Since the functions of a product can increasingly be scaled and modularised, a demonstration with a technological functional prototype is no longer expedient. The sub-project called "Blended Prototyping" examined how iterative user tests with prototypes can help to build a bridge between different levels of complexity in development. Increasingly, product development involves holistic systems with manufacturing and service components, infrastructure and business models. The "Hybrid Prototyping" sub-project answered the question of how these systems can be tested in a user-centred way. All three sub-projects sought hybrid prototyping approaches in areas in which no longer the complementary use of the technological and design prototype, but rather their merging could lead to a holistic solution.

The jointly undertaken attempt to differentiate between model and prototype also confirms the blurred boundary between the previously co-existing terms. According to the traditional point of view, the model was upheld as a simplified or reduced, but primarily theoretical and abstract replica of a complex reality, yet one which represents, in its objectified form, a fluid transition to the prototype. Above all, this applies to the generative models of designers that approach the complexity of reality in constant change and thus become prototypes. By contrast, the prototype represents a higher degree of complexity (in regard to the specific issue), which is why it can fulfil concrete demands, be tested, validated, verified and evaluated.

From the point of view of the engineering sciences, models and prototypes traditionally do not embody the degree of abstraction that is present in the generation of ideas, but rather a degree of maturity with respect to the realised draft in the sense of an analytical consideration aimed at a pre-defined goal. The levels of development in this sense trace the course of idea-model-demonstrator-prototype-product, with the model understood as the very general first materialisation of the idea. In common parlance, the prototype is considered proof of the correctness of an idea or objective, and can be understood as the first archetype of the product. This project has showed that not only a flowing transition, but rather also jumps are seen in this area: A model (even "just" a sketched-out idea) turns into a prototype by means of rapid prototyping, which can directly be the finished product as the "Beyond Prototyping" sub-research project showed. Quick production methods and algorithmically-generative and digital tools allow an approach to areas of engineering and artistic-design disciplines that initially think and operate in a complementary way.

At the end of the clarification processes for the terms, all the participants were in agreement that the role of the prototype as medium would be recorded as the smallest common denominator among the prototyping concepts. In terms of specific issues, a prototype is a mediating element between the actors involved in it. Prototyping processes are therefore at the core of communication processes.

3 From Static Prototypes to Dynamic Prototyping

The considerations on discernible dichotomies and the attempts at defining a prototype led to the following discovery: A view of the prototype that statically reflects a specific stage of development is not expedient in a holistic consideration of development processes in which the actors must fulfil systematic requirements in a multi-competent team. Therefore, the research group conducted prototyping processes in mixed groups in a prototyping workshop, and dedicated themselves to the analysis of development processes. Multiple development tasks from distinct fields were addressed jointly and documented on different levels (e.g. procedural and terminological). In the subsequent evaluation of the work and communication processes, it was possible to develop a discipline-overarching description of the prototyping process (see Fig. 1).

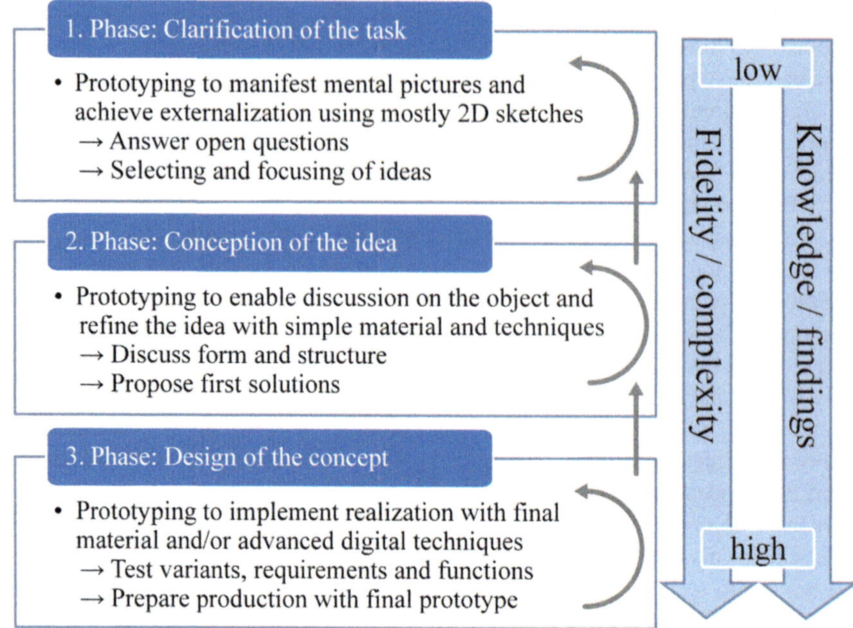

Fig. 1 Prototyping process (Exner et al. 2015)

The combination in this description was by no means trivial since nine disciplines (Digital Design, Industrial Engineering, Mechanical Engineering, Architecture, Automotive Engineering, Interaction Design, Computer Science, Cultural Science and Physics) were involved in the process and they brought greatly diverging views and focal points along with them (Exner et al. 2015). This abstract description of the prototyping process is integrative and represents a basis for communication in transdisciplinary development teams. The integration of distinct dimensions into the abstraction of an ideal-typical prototyping process helped achieve greater penetration than has been seen in the conventional, very generic attempts at definition.

Another step was the attempt to derive a collective prototyping definition from the workshop results. Although the distinctly used terms such as drawing, mock up, draft, simulation, model, etc. could ultimately be identified as partial aspects of a holistically observed prototyping term, a definitive, collective definition was ultimately not possible. Instead, this attempt raised the question of whether a complete description of applicable characteristics of prototyping across disciplines in the form of a definition is expedient. The research group considered it more sensible to describe the individual prototyping approaches according to their functional focal points (communication, validation, determining user perspective and generating ideas) and to position them in a multi-dimensional matrix. This facilitates a differentiated description of prototyping processes across disciplines and thus communication on their diverse potential and the resulting possibilities for use in

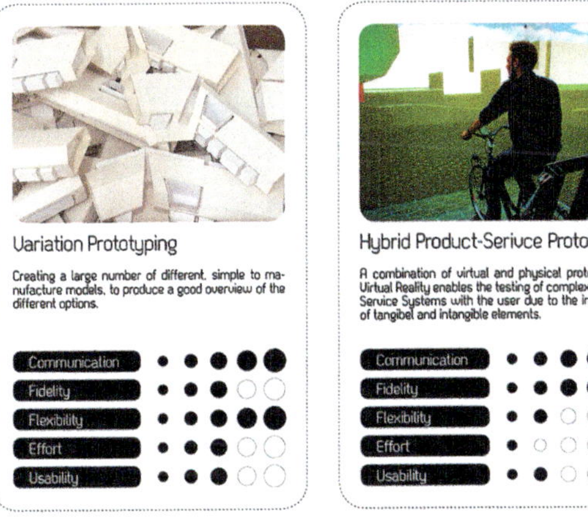

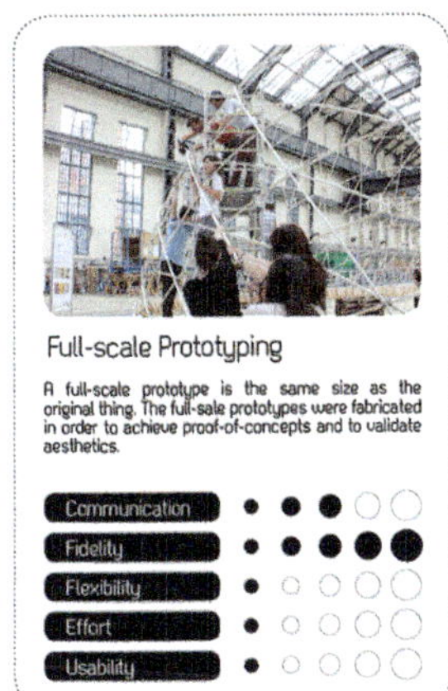

Fig. 2 Prototyping methods/prototyping quartet

interdisciplinary teams. The dimensions of the created matrix are the aspects of prototyping known to all the disciplines involved in the project: effort, fidelity, flexibility, usability and communication (Fig. 2).

The prototyping workshop, where these considerations and discoveries were addressed, was a valuable format for this project in order to question one's own perspective and to enrich the discipline-specific ways of thinking and procedures by obtaining ideas from other disciplines. The findings in the workshop, particularly the idea of a clear positioning and integrating individual prototyping concepts in a matrix, were practically implemented in the prototyping app for smartphones and in the prototyping quartet (Fig. 2). Both are elements of the expanded final publication for this project, the *layer cake*.

4 Layer Cake

In the three-year research process, the sub-projects developed the hybrid prototyping concepts that were presented in the preceding chapters of this anthology. The research group's objective was also to depict the research results in an integrative form that corresponds to the research principle of transdisciplinarity rather than to arrange them additively alongside each other in a standard collection (cf. Eichmann and Nagy in this volume). In addition to the prototypical self-reflecting, self-optimising project, an appropriate format was developed that reflects the character of the research. It is a package that includes this book and other artefacts layered one on top of the other. These layers transport the findings according to the principle of understanding by doing, which reflected a central aspect alongside the theoretical considerations in this research project. This form of publication offers access to research contents on multiple levels, so-called *layers*, and therefore

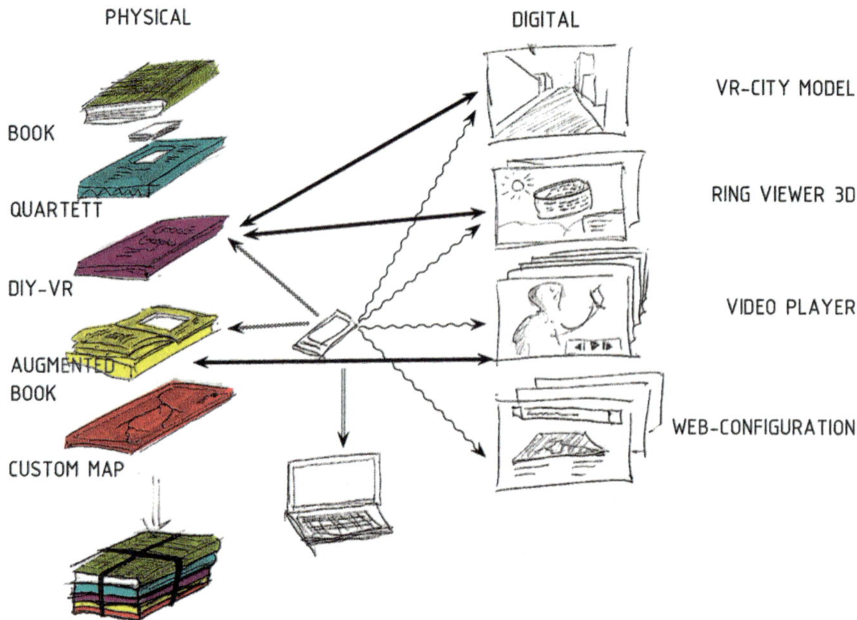

Fig. 3 Structure of the layer cake

abandons the framework and linearity of a book. The analogue and digital elements of this so-called *layer cake* impart knowledge from the considerations on the joint understanding of prototyping in a playful, appealing and generally understandable way. Figure 3 illustrates the structure and the concept of the publication of all the research results, which the research group understands as a prototype of a transdisciplinary publication.

The book, as the top layer, contains all of the scientific and academic findings from the individual sub-projects, the overall project and the projects accompanying this transdisciplinary project. The design of the cover reflects the increasing relevance of individualisation in product development: The regular pattern provides a scaffold for the owner of the book to customise its message. The cover of the book is inspired by random international's work in the early 2000s (http://random-international.com/work/tape/).

Quartet and Prototyping App The prototyping quartet card game and the prototyping app, as already described above, playfully reflect the results of the discussions on prototypes and prototyping in the search for a general understanding of prototyping. Prototyping quartet consists of 25 cards that show prototyping methods and the evaluation of prototyping properties. The description of the methods is carried out with the help of five categories (effort, fidelity, flexibility, usability, communication). Each category can be evaluated with a maximum of five points. The ratings can help in selecting the right prototyping methods. Comparing and displaying the quartet cards will amusingly introduce a player introduced to the topic of prototyping. The contents are compiled by all project participants and are also used for the prototyping app, which is part of the applications within the augmented book. The prototyping app allows a development team to select suitable prototyping methods at different phases of the development process. Its interface also makes it possible to evaluate distinct factors such as expenses or communication. Finally, the app recommends multiple prototyping processes in a list. In addition, the user receives information and examples on how to proceed. The interface also offers the option of including additional prototyping processes in the app, along with their advantages and disadvantages, and thus places them at the disposal of users. The app thereby supports the search for new prototyping possibilities and makes it easier to try out different procedures.

Do it Yourself Virtual Reality For a better understanding of what virtual reality is and why immersion, interactivity and the human imagination are so important, we have built a simple prototype based on the Google Cardboard project. Google Cardboard is a simple HMD (head mounted display) consisting of a cardboard, two wide-angled lenses and a smartphone. Our approach is to empower the reader to build his or her own HMD prototype. For this reason, the project team prepared a cardboard and one virtual reality application, which is a ready-to-use smartphone. Following the instructions, the reader is able to build, see and understand how virtual reality and HMD displays work: The application makes it possible to place aspects of individual sub-projects at the disposal of users in an exploratory way. Accordingly, digital models from “Hybrid Prototyping”, such as the digital city model of Berlin or the Pedelec product are visualised and explored by the user (cf. Exner et al. in this volume). Additionally, “Beyond Prototyping” enables a virtual previewing of an instance of the *Ciphering* (cf. Ängeslevä et al. “Beyond Prototyping” in this volume), enabling the user to align the model to decode the hardcoded message in the ring. Besides an additional display of the project results, which cannot be explained in a book, the complexity of virtual reality is reduced and thus made possible for the user to experience in a playful way, which is one of the main characteristics of prototyping.

Augmented Book The layer-augmented book creates a self-made book that integrates physical book pages with interactive content on mobile devices, inspired by the Kickstarter project “Little Magic Books”. The book uses a mobile device that is attached on top of the last page. Through cutout areas in the other pages, a reader/user can see and interact with the device’s display. After a specific app is installed on the device, it automatically detects which page the reader opened. This is done with small metal feelers that are integrated on the backside of the pages and trigger touch commands on the device display.

In this way, the device can provide content to the user that is related to the specific physical book pages. For such content, we use videos, 3D models, diagrams, photo galleries and a special medium—a film about the entire “Rethinking Prototyping” project. Furthermore, with touch gestures that bridge the space between the physical page and device display, the user can make references from the content printed on the page to the app installed on the device.

Custom Map Locatable is leveraging the social context that tables can provide and bringing meaningful and aesthetic customisation to "the table" (cf. Ängeslevä et al. "Beyond Prototyping" in this volume). As a layer cake component, a whole table is not feasible, and therefore an instance of locatable is produced that is ambiguous in its use. The chosen area of the map depicts the partners in the "Rethinking Prototyping" project, potentially serving as a talking point for the partners involved in the project.

5 Conclusion

The openness to questions and results in the project made it possible to flexibly circumvent the initially set goal of a collective definition of prototyping. The project's first results from comprehensive discourse raised the question of the extent to which a joint and holistic definition of prototyping can and should be sought at all. It was determined that it would be more expedient to work out fundamental factors that are applicable across disciplines and which, when transferred to a matrix, allow for a clear positioning and description of individual methods and concepts. Finally, it is important to note that the findings from the joint work on an overarching understanding of prototyping that produced the matrix concept, may only be preliminary at the present time. Nevertheless, they are also ground-breaking. The individual dimensions that are used to identify the various prototyping concepts across disciplines by their position in the matrix present an opportunity for future research that will theoretically justify and describe these in greater detail.

In addition to the theoretical findings, the individual sub-research demonstrated interfaces between the different disciplines and their concepts, which were subsequently used as a starting point for the development of concrete hybrid prototyping concepts. The differentiating, but also integrating consideration of prototyping in the theoretical discourse as well as the development and testing of hybrid prototyping concepts in practice facilitated the productive, transdisciplinary work on the research subject matter. With respect to the methods, processes, functions, areas of application and the objectives of the prototyping and prototypes, it was possible to achieve a more in-depth mutual understanding between the individual disciplines. On the basis of the experiences and findings in the project, such an understanding on a broader level can also be understood as a general prerequisite for the holistic development of new products and complex, interactive systems.

Reference

Exner, K., Änglesleva, J., Bähr, B., Nagy, E., Lindow, K., & Stark, R. (2015) A transdisciplinary perspective on prototyping. In *International conference on engineering, technology and innovation* (Belfast, Ireland, 2015). Institute of Electrical and Electronics Engineers (in press).

Part III
Joint Research

Reflections on Transdisciplinary Research

Ulrike Eichmann and Emilia Nagy

Abstract In this chapter, the project coordinators reflect retrospectively upon the most important elements of the transdisciplinary collaboration in the "Rethinking Prototyping" project. On the macro level, the fundamental importance of reflective-coordinating support is outlined against the backdrop of ambivalent exp eriences with inter-/transdisciplinary research, and the assumed added value of transdisciplinary research for this project—the integration of knowledge—is described. A general overview provides the challenges within the science system that reflecting-moderating support of transdisciplinary processes must address in various ways, depending on the project. With recourse to project-internal documentation, empirical values and the results of an accompanying study, the most important elements of the collaboration are then elucidated on the micro level and assessed with regard to their potential for the promotion of the process of knowledge integration. Based on the results of this evaluation, beneficial factors for knowledge-integration and transdisciplinary collaboration are worked out. Throughout the course of the project, the guiding principle that each transdisciplinary project is unique and must be understood as prototypical was developed. Transdisciplinary projects are implemented in the form of a continuous development process that, as summarised at the end, is to be understood as part of a global prototyping process in transdisciplinary research. This paper makes a contribution to this subject.

1 Introduction

The technologically-produced complexity of our world is increasingly penetrating into every sphere of our lives. Also intertwined and interdependent are the questions, challenges and problems that this world gives rise to. The answers and

U. Eichmann · E. Nagy (✉)
Hybrid Plattform, Berlin University of the Arts, Berlin, Germany
e-mail: emilia.nagy@hybrid-plattform.org

U. Eichmann
e-mail: ulrike.eichmann@hybrid-plattform.org

C. Gengnagel et al. (eds.), *Rethink! Prototyping*,
DOI 10.1007/978-3-319-24439-6_12

solutions required for this must do justice to this complexity and therefore also be developed systematically and in a context-related way according to a holistic approach. For science, this consequently requires a research principle exceeding the limits of individual disciplines. With the simultaneous participation of multiple scientific and artistic-creative disciplines as well as society, new approaches must be sought, innovative solutions should be developed and new knowledge produced. Although the necessity of transdisciplinary research is demonstrated in this, it is not self-explanatory or simple to carry out inter- or transdisciplinary projects. For example, in order to go beyond the limits of disciplines, institutes, universities and non-academic establishments that generally work in isolation, it is necessary to have additional time, staff resources and financing as well as a special infrastructure.

The research project entitled "Rethinking Prototyping" was able to overcome many adversities that transdisciplinary research is exposed to in the university context, but had to prove itself in the implementation of some challenges. This paper has been written from the perspective of the project coordinators[1] in the research group of "Rethinking Prototyping" and addresses observations and knowledge gained from supporting this project. The focus is on the factors that are related to achieving *knowledge integration* that leads in in an ideal case to the achievement and/or answering of the transdisciplinary goal/question as well as to new knowledge and assessments by the individual participant. These results can have an impact not only in the project context, but also in the respective disciplines due to the participants' use of them.

In the following, the focus is more on the framework of organisation and support that influence the integration of knowledge. We describe *formats* and *design elements* that were used in order to successfully pave the way for the transdisciplinary research process on the level of the organisation of collaboration in a *coordinated, integrated, supportive, advisory* and *facilitating* way. These explanations can serve as sources of inspiration and orientation aid for conducting other transdisciplinary research projects.

The basis of this chapter is the idea that each transdisciplinary project can be understood as a prototyping process of transdisciplinary research. Consequently, we do not assume that there is *one* right way in transdisciplinary collaboration. In this sense, no normative claims are made in this chapter, nor is a final evaluation of the research project provided. This paper refers to one single transdisciplinary research project involving the Technische Universität Berlin (TU Berlin) and the Berlin University of the Arts (UdK Berlin), reflecting on the work in this specific university context and considering the extent to which it can serve as a model.

The basis for this paper consists of the documentation of the course of the project, the work meetings, accompanying research[2] and the observations and

[1]The project coordinators consisted of a project manger and a project administrator, who worked together closely on the conceptual level during the course of the project.

[2]The dialogic and process-accompanying research was conducted by Maria Oppen from the Social Science Research Center Berlin (Wissenschaftszentrum Berlin für Sozialforschung/WZB) on behalf of the "Hybrid Plattform" from January to December 2013. This involved an accompanying

analyses of the project coordinators. To start with, a theoretical framework is provided on the macro level (Sect. 2): the transdisciplinarity term adopted for this project is explained, the concept of knowledge integration is specified, and the levels of influencing the design of knowledge integration are outlined. This is followed by an enumeration of the possible challenges which transdisciplinary research may potentially face in the university context and which can influence the intensity of the collaboration and thus integration of knowledge (Sect. 3). Then, on the micro level, i.e. on the level of the organisation and design of the concrete "transdisciplinary scientific practice" (Balsiger 2005, 170), the formats used in the service of knowledge integration are analysed retrospectively and the factors for success in the achieved integration of knowledge in the "Rethinking Prototyping" project are worked out (Sect. 4). In conclusion (Sect. 5), we summarise our most important experiences and discoveries.

2 Theoretical Framework

2.1 Inter-/Transdisciplinarity: An Ambivalent Phenomenon

It is not easy to fully develop the potential of interdisciplinary research in practice: "Based on their own extensive experience in research, various authors have described interdisciplinarity as an ambivalent phenomenon" (Laitko 2011, 1). The same applies to transdisciplinary projects. This is because the expectation that extraordinary results will be achieved by bringing together various disciplines cannot be easily met under even the best conditions[3] (cf. Laitko 2011, 9f.). When the limits of a discipline are exceeded, scientists enter an area in which they are often confronted with unusual or unfamiliar processes that differ from project to project. In an ideal case scenario, project participants have the will and motivation to work across disciplines, but they can rarely fall back on familiar or established procedures. This situation has been the cause of ambivalent experiences. The comments made by one of our project participants demonstrated this, for example. He said that most scientists wanted to research across disciplines, but nobody could resolve the difficulties to an adequate extent, although they are all well known. The ambivalence in transdisciplinary research is due to the fact that the high expectations for inter-/transdisciplinary research on the macro level are difficult to fulfil on

---

(Footnote 2 continued)

process evaluation of the work for the "Hybrid Plattform" in the context of which the "Rethinking Prototyping" research project was included and examined. The scientist briefed the project coordinators multiple times, interviewed eight project participants and presented the intermediate results of her research to the entire research group within the context of a large colloquium. The results were published in 2014.

[3]Hubert Laitko analyses the history of the Starnberg Max Planck Institute for Research on Living Conditions in a Scientific-Technical World.

the micro level of university research because there are (still) no successful realisation strategies for implementation. The success of transdisciplinary projects is often random. The initiators hope that the project participants will provide knowledge of possible work and organisational forms for functioning collaboration in the group and can initiate and maintain knowledge-generating processes. But this is frequently not the case. This often produces confusion and irritation and has a demotivating impact on the participants (cf. Schmithals et al. 2011, 28, 56).

The ambivalent experiences that result from the divergence between the high expectations and the lacking realisation strategies show that successful collaboration does not function or only rarely functions by itself within an trans-/interdisciplinary research group. This is also what Gert Dressel et al. say: "Inter- or transdisciplinary research is not without conditions, it does not happen by itself, but rather must be organised systematically" (Dressel et al. 2014, 207). We see the need to accompany inter-/transdisciplinary research processes in a coordinating and supportive way. The outlined phenomena for the discrepancy between high expectations and the (still) lacking realisation strategies for inter-/transdisciplinary research should be countered with reflective processes and suitable formats in order to develop and exhaust the desired added value of inter-/transdisciplinary research.

2.2 *Potential of Transdisciplinary Research: Knowledge Integration and Self-Renewal of the Disciplines*

If we return to Jürgen Mittelstraß's definition of transdisciplinarity, we can see the added value that the "Rethinking Prototyping" research project pursued with its transdisciplinary approach. Mittelstraß argues that there is a need to go beyond disciplinary limits in (at least) two factors on the macro level. On the one hand, individual disciplines could no longer provide comprehensive answers to the growing complexity of problems in everyday life (cf. Mittelstraß 2003, 8). On the other hand, disciplinary research benefits in terms of innovation since new knowledge arises "on the edges, between various subjects and disciplines and in their connection to each other" (Mittelstraß 2008, 5).

For our understanding of transdisciplinarity, its contextualisation in application-oriented research plays a subordinate role although there was reference to practice in the sub-projects. According to our understanding, "transdisciplinary research is not application-oriented per se" (Schmithals et al. 2011, 46), but rather we observe its innovativeness in terms of the development of new knowledge and the related self-renewal power of disciplines as its primary quality and function. Its innovativeness is defined primarily in the project context itself: New knowledge is produced through collaboration with the project participants. This internalised knowledge and the experiences of the participants also reflect back on the individually-involved disciplines; the power of self renewal of the disciplines provoked by transdisciplinary research has an impact here. In the following, we concentrate on how new knowledge arises in a specific transdisciplinary connection and what can be described primarily through the process of *knowledge integration* according to our approach.

The concepts of multi-, inter- and transdisciplinarity demonstrate what is meant by *knowledge integration* in our analysis. We shall characterise these three forms of cross-disciplinary practice, starting with their desire to integrate knowledge. In the process, we fall back on the three levels of integration according to Günter Ropohl: *encyclopaedic integration*, *interpersonal integration* and *intrapersonal integration* (cf. Ropohl 2010, 4f.).

Multidisciplinarity is characterised by no integration of knowledge or a minimal amount. In multidisciplinary constellations, only the level of *encyclopaedic integration* is achieved. This approach collects "the important disciplinary perspectives in an additive way" (Ropohl 2010, 4f.; cf. Laitko 2011, 11) and does not require any collaboration on a collective issue. The results of multidisciplinary research are usually included in collections in the form of individual papers and are "arranged without theoretical interconnections and in an unrelated way" (Ropohl 2010, 4f.).

By contrast, the term *synthesis* describes the sought degree of knowledge processing for *interdisciplinarity*. Various approaches should "merge" into a collective answer to a research question (cf. Ropohl 2010, 4f.). Furthermore, Ropohl explains:

> If the results of the work [...] should go beyond being an aggregate of specialised expertise, the participants must have good communication skills and a strong ability to learn in order to synthesise their individual contributions (Ropohl 2010, 4f).

Interdisciplinarity requires a joint research question, learning and communication skills and finally a synthesis. Ropohl calls this form of integration *interpersonal integration*. He notes critically in this regard that the results of interdisciplinary research frequently only achieve the level of encyclopaedic integration, so the research remains, if defined strictly, multidisciplinary. He views the reason for this as being the lack of "methodological tools" and suitable competencies (cf. Ropohl 2010, 4f.).

Transdisciplinarity refers to interdisciplinarity with a completed act of integration—or as Mittelstraß puts it:

> Interdisciplinarity in a correctly understood sense does not move between disciplines or hover, like the absolute spirit, over the fields and disciplines. Rather, it eliminates disciplinary narrowness where this stands in the way of the development of the problem and corresponding research action: speaking accurately, it is transdisciplinarity (Mittelstraß 2003, 9).

Accordingly, interdisciplinarity, in a falsely understood sense according to Mittelstraß and without interpersonal integration according to Ropohl, is simply multidisciplinarity. Following this interpretation, the term interdisciplinarity becomes superfluous (see Fig. 1) and is described here as transdisciplinarity for the project work of "Rethinking Prototyping".

The degree of knowledge integration in transdisciplinary research can be explained on the level of *intrapersonal integration* according to Ropohl (2010, 5). This requires, according to our interpretation of Ropohl, not only the aspects for

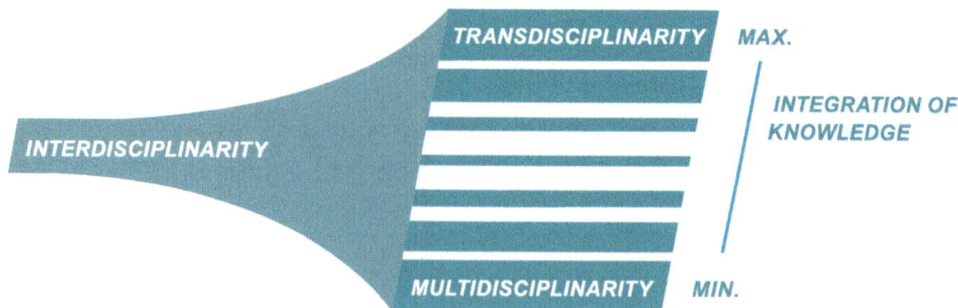

Fig. 1 The level of knowledge integration defines the character of the joint research ranging between multi- and transdisciplinarity

interpersonal integration (willingness to exchange and communicate in order to produce a synthesis), but also "individual multi-field competency" (Ropohl 2010, 5) from the project participants. For Ropohl, this includes the ability of the participants to be able to understand and integrate knowledge from the widest range of disciplinary origins on an individual, i.e. intrapersonal level. Intrapersonal integration also means that "this person passes on the synthesis of knowledge not only receptively in him- or herself, but also effectively to others" (Ropohl 2010, 5).

Building on Ropohl's definition, according to our interpretation, transdisciplinary projects pursue the goal of answering a question through synthesis of the knowledge available in the project, which is continuously renewed and changed through individually completed integration processes. The integration of knowledge, as we understand it, means that project participants record new knowledge and new methods on the intrapersonal level, integrate them into an existing body of knowledge and gain new knowledge, new processes of knowledge attainment and new discoveries through the analytical processes of differentiation and synthesis. On the individual level, it is a critical-reflective absorbing and understanding of other perspectives in the reflection of one's own body of knowledge and requires a certain willingness to revise and expand one's own perspectives. The discoveries made in this way are repeated and "thought-through" for their potential by the other project participants. The integration of knowledge is thus understood as a circular-dialogic process that runs like a spiral and leads to the answering of a joint question in this way. This process of handling knowledge has an effect, both within a project and beyond its limits. The intrapersonally processed, newly attained knowledge flows back into the respective disciplines through the project participants. Therefore, knowledge integration processes also stimulate a circulation of knowledge between a transdisciplinary project and its involved disciplines and institutions, which may be inspired or changed as a result of this.

The possibilities for promoting the difficult and multi-layered integration of knowledge are heavily influenced by the specific framework conditions in the scientific system; this has been confirmed by our experiences and the results of the dialogic research that has accompanied the "Rethinking Prototyping" project. In order to pave the way to knowledge integration, it is important both for the participants and on the coordination level to develop awareness for these framework conditions in order to determine the room to manoeuvre. In the following, we shall address the most important influential factors in the scientific system for transdisciplinary projects.

3 Context-Related Challenges: Finding Room to Manoeuvre

Every scientist involved in a transdisciplinary project is embedded in an environment that consists of systems with higher-level goals and values.[4] In this section, we will examine some examples of factors that (a) determine the leeway both for individual project participants and for reflecting-moderating support and (b) can influence the process of knowledge integration. These factors were worked out and analysed for the most part in the research accompanying the "Rethinking Prototyping" project (cf. Oppen and Müller 2014, 38–46). They will be complemented here by the experiences and observations of the coordinators in this project. These explanations should provide initial clarification in general of the limited options for action in project support and design as well as the incomplete degree of freedom that the participants have in a transdisciplinary project not isolated from external influences. They create the framework for the subsequent project-specific explanations (Sect. 4) that allow for reflection upon the most important elements in the collaboration on this project and an assessment of their potential for the promotion of knowledge integration.

The first influential factor that has an impact in transdisciplinary projects can be called the international scientific system. Its influence extends from its subsystems according to the subsidiarity principle to the micro level of a research project. The zeitgeist of international scientific work (e.g., trending subjects that are rewarded with great attention and funding) can influence, for example, the formulation of a research question or the motivation of individuals independently of their disciplinary affiliation.

The subsystems of the scientific system, as well as the disciplines, universities and research facilities with their departments and institutes have an impact on individual scientists not only during their scientific career, but also in the course of a transdisciplinary research project. Accordingly, scientists are shaped by their discipline, for example, through their theoretical background and the language in their field, and bring a specific internalised discipline culture into the project. In the course of their socialisation in their respective discipline, scientists acquire a specific intellectual and research culture that can be juxtaposed diametrically in a transdisciplinary group such as, for example, quick focussing versus a cautious approach, linear causal models versus non-linear creative theoretical approaches,

[4]Based on Talcott Parsons' theory of social systems (cf. Stark 2009).

specific versus holistic analysis, risk willingness versus planning security, discussions with a change in perspective versus work in isolation, creative freedom and individual design versus meticulous planning and hierarchically-controlled project organisation (cf. Oppen and Müller 2014, 42ff.). The different degree of compatibility in these theoretical and communication patterns can influence the working atmosphere in a project. Project participants differ in their willingness and ability to be aware of one's own internalised patterns and to expand or revise them. Usually, participants must be open to new work methods. Institute-specific methods can assume a place of "sovereignty" in a project if the place of research is primarily tied to only one location, for example.

In transdisciplinary collaboration, the different discipline cultures become very evident and can lead to implicit hierarchies within the research group. Accordingly, project participants in a transdisciplinary framework can be mapped in a hierarchical structure that corresponds to the disciplines and is also taken for granted in the project. Prejudices with respect to other disciplines or disciplinary stereotyping play a significant role here (cf. Oppen and Müller 2014, 42ff.).

Furthermore, differences between documentation cultures and knowledge management present a challenge for the accessibility of the available knowledge within a project. Accordingly, for example, there is the danger that project-relevant knowledge remains in the archives of the individual institutes. Limitations in the exchange of data on account of data protection requirements can also cause complications in the continuous flow of information and the exchange of knowledge between project participants. Furthermore, the respective organisation structures, communication forms, management cultures and control mechanisms in an institute (e.g., hierarchical or democratic) should be named here as factors. They determine, in particular, the exchange of information between professors and research assistants. In this regard, the quantity and quality of the project participants' knowledge input differs significantly, which also causes the participating disciplines to have a different presence.

Last but not least, there are the involved scientists who can themselves determine their own room to manoeuvre within a certain framework and thus also influence the coordinating-reflective support in transdisciplinary research projects. Accordingly, each scientist pursues individual goals with respect to his or her activity in the scientific system (cf. Oppen and Müller 2014, 40). If the focal points of the project participants' research is more in the core research areas of the discipline, transdisciplinary research is less conducive for the given scientist's own interests since transdisciplinary research questions usually only relate to the disciplinary questions to a limited extent. The success of a transdisciplinary project can also depend on the extent to which the individual participants view the collectively-achieved transdisciplinary collaboration as useful for themselves. The feedback of the knowledge integration into the participant's disciplines can affect the fact that the disciplinary assessment of the collective, transdisciplinary question is viewed positively if the developments and results of the transdisciplinary group are also relevant for the discipline. In order to strengthen the integration of the knowledge between the project and the disciplines, it appears sensible, for example, "to relate the results of an interdisciplinary project to general questions and problems in the individual disciplines" (Arnold et al. 2014, 117). This strategy makes it possible to also pursue personal goals that may have a stronger disciplinary focus within the framework of a transdisciplinary "affair".

This section described some central influential factors that determine the course of a transdisciplinary project and the leeway in the promotion of knowledge integration. It is clear that the challenges resulting from them cannot be considered in full for designing the process of a transdisciplinary project and cannot always be successfully encountered. Awareness of these factors is, however, indispensable for an assessment of the degree of freedom that project participants have in transdisciplinary work and the actually available range for reflecting and coordinating support. In the "Rethinking Prototyping" project, they were constantly reflected upon and considered in the realisation of the research project. Against this backdrop, the following experiences from the implementation of this project shall be evaluated.

4 Transdisciplinary Research Elements in the "Rethinking Prototyping" Project

In addition to the previously outlined, generally systematic and actor-based factors for transdisciplinary collaboration, the specific realisation of the "Rethinking Prototyping" project will now be described here in more detail, particularly with regard to the formats and elements of collaboration that were used for the support and promotion of the transdisciplinary integration of knowledge. Initially, the fundamental project structure will be explained. It reflects specific framework conditions under which the implementation of the transdisciplinary collaboration was to be achieved in this project. Then the formats and elements of collaboration are illustrated in their form and realisation, and their effect on the integration of knowledge is assessed.

4.1 Basic Structure

"Rethinking Prototyping" was the first project carried out by the TU Berlin and the UdK Berlin on their joint transdisciplinary "Hybrid Plattform".[5] Transdisciplinary collaboration between various disciplines at the two universities was achieved on two levels in the "Rethinking Prototyping" project.

In each of the three sub-projects, research assistants[6] from at least two fields at the UdK Berlin and the TU Berlin worked under the lead of at least one professor at

5Cf. introduction to this book and the platform www.hybrid-plattform.org for more information.

6The term "research assistant" is understood to be the engagement of the involved architects, designers, softwaredevelopers and engineers.

each university and addressed individual aspects of the prototyping (sub-project level).[7] Furthermore, the overarching object of research was the question of whether there is a joint concept of prototyping (overall project level) (see Fig. 2). This question was formulated during the concept and application phase as, so to say, the *objective* and the *basis* of the joint research on the limits of the disciplines. The required agreement on the meaning of central terms, methods and concepts was tied on the one hand to theoretical discourse with the goal of defining the term prototyping. On the other, this theoretical-methodological reflection was understood as an opportunity to gain new ideas for research in disciplines that are in part not related to each other. In the initial project application concept, epistemological expectations were defined for the involved scientists and designers, but primarily application-oriented disciplines were represented in the project. Besides the claim to theoretical discoveries, there was also great interest in practical solutions.[8]

In the project application, coordination was planned to support the project by working with the heads of the project and closely collaborating with the research group. The project coordination level represented the organisational framework in order to determine and support the collaboration in the terms of theoretical and practical knowledge interests and to promote the integration of knowledge both on the overall project level and at the interface to the sub-project level (see Fig. 2). At the kick-off meeting to start the project, the formats for collaboration in the sub-projects and for all the participants, as set forth in the application, were specified for the entire course of the project, and their implementation planned. The formats were partially handled in a flexible way during the course of the project and successively adjusted to the existing needs in the project. Methodological impulses and offered formats for knowledge-integrating cooperation between the sub-projects on the overall project level were primarily developed and implemented by the coordinators (external organisation), but also came from the project participants (self-organisation), which increased accuracy and acceptance within the group. The desire for self-organisation required that the project participants address not only purely content collaboration, but also organisational-methodological issues in transdisciplinary collaboration, which consequently also made up a focal point of the joint meetings.

In designing the research process, the coordinators (on the coordination level) were always dependent on the participants' consent and the willingness to act. A particular challenge also consisted in the fact that not all the project participants were equally involved on the overall project level. The intensive exchange on the joint research question primarily took place on the level of the research assistants. They were subject to the instructions of the professors who were more heavily involved in the sub-projects. To make sure these instructions were in the interests of

---

[7]Cf. sub-project results from "Hybrid Prototyping", "Blended Prototyping" and "Beyond Prototyping" in Part II of this book.

[8]Cf. the system-theoretical analysis of the engineering sciences for more on this, additional information in Ropohl (2010).

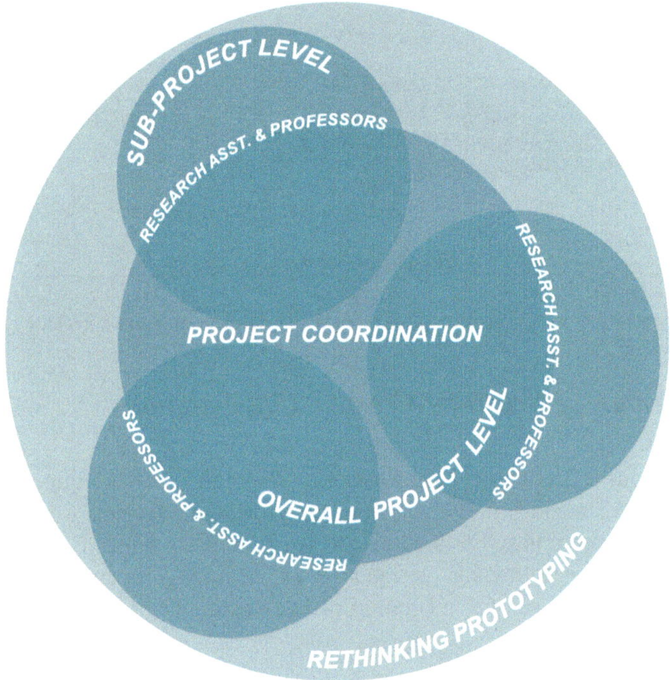

Fig. 2 Basic structure of the project "Rethinking Prototyping"

the project as a whole, the professor level was regularly informed about the integration processes on the overall level. This made it possible to simultaneously expand the circulation of knowledge integration to the involved institutes.

The parallel guidance of research collaboration and the methodological-organisational procedure with consideration given to the relationship between the overall project level and the sub-project level meant that the coordinators also had to take into account these supporting organisational frameworks. The cooperation between the research work level and the coordination level had to be balanced out over the course of the project and mutual expectations had to be clarified so that there was no confusion in terms of the respective roles and responsibilities in externally-determined and self-determined matters with regard to the form and content in the transdisciplinary research. The collaborative work between the two levels was developed in integrative collaboration that proved retrospectively to be very constructive for the integration of knowledge.

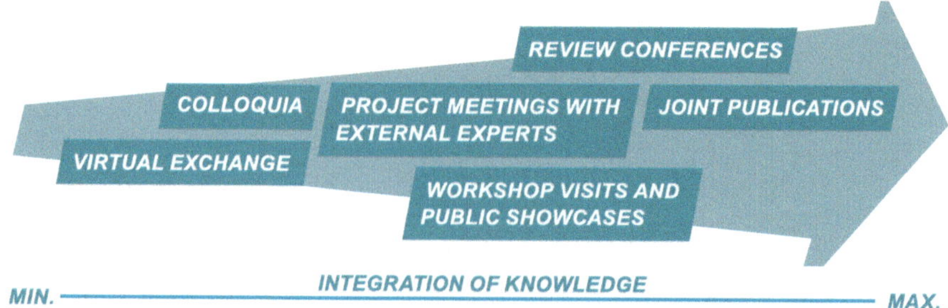

Fig. 3 Formats used in this research project respective to their effect on knowledge integration

4.2 Formats of Scientific-Creative Collaboration

The continuous reflection of the formats of collaboration was an important part of the coordinating support. In the course of the project, the formats were adjusted or supplemented on the basis of meta-discourses in order to intensify the integration of knowledge. The following presentation of the used formats is based on the degree of their effect on the integration of knowledge, beginning with the lowest (see Fig. 3).

4.2.1 Virtual Exchange

In projects across multiple institutions, the use of a web-based cooperation platform is important for internal collaboration. This is because it is possible to add and intensify the exchange of content on a virtual level, which makes organisation and documentation easier. For collaboration in the "Rethinking Prototyping" project, a co-working platform was set up after approval was given in the project group.

This encouraged the exchange of content by creating discussion forums for central terms in the project, for example.[9] Creating a comprehensive list of literature was also possible on the joint work platform. Furthermore, joint projects such as teaching events were also planned and subsequently addressed via the platform. The project group also virtually organised parts of the joint work process on the publication concept. For example, it developed and collected collective content for a prototyping quartet card game. The co-working platform also acted as a knowledge archive; the project meetings were documented and the work steps and results were recorded for the project participants. This happened, for example, via protocols and audio recordings of joint meetings or the archiving of presentations on the latest status of the sub-projects with brief summaries of the results from the discussions.

[9]It also initiated discussions of fundamental and higher-level questions such as "What does prototyping mean for you?" and encouraged the clustering of the results on the online platform.

In the course of the project, the co-working platform was viewed increasingly and, toward the end, primarily in its function as a knowledge archive and used for the documentation of the project through coordination. In regard to the potential for the platform to be a knowledge base for the integration of knowledge, the following retrospective challenges had to be addressed in the project.

The successful implementation of such a co-working platform assumes that its functions and its use are considered collectively as useful and it meets with broad acceptance in the team, and the team members are willing to use it. This was only the case to a limited extent in the project. We see the following reasons for this: In the actual course of this project, the theoretical discussions shifted somewhat, away from purely verbal exchange and increasingly toward the practical area of the joint prototyping, which functioned via personal presence. In the course of this shift, we saw the limits of the virtual co-working platform, which was less effective for this practice-based discourse on joint research questions and was also used less as a result. The documentation of the findings and results that the participants achieved in practice through their experiences would have required their linguistic or visual preparation. However, the participants' capacities and also their general willingness was lacking in the project.

In order to make sensible additions to the already existing, but separate institute-specific documentation structures (cf. Sect. 3), continuous use of the platform by all the project participants would have been important in order to guarantee completeness in content, for example. But it was difficult to establish this continuity since the consistently present capacity of the individuals was lacking for the updating of a double documentation structure (institutional and project-related). It was seen, however, that the motivation to use the platform was high in phases when joint work steps or joint projects were started (e.g. review conferences, joint public events, colloquia), but it remained sporadic and levelled off shortly afterwards. Knowledge documentation did not take place to a complete extent as a result.

Despite our only limited positive experiences, we are of the opinion that it is necessary to establish a joint level of knowledge management for transdisciplinary projects in order to virtually add processes of knowledge integration through this knowledge management. It is important to include the use of such platforms from the beginning in the planning of transdisciplinary projects and to set them up shortly after the beginning of the project. It should also be stressed that the sum of the individual, partially institution-related documentation (encyclopaedic integration) without integrative moments can still produce no knowledge-integrating comprehensive documentation. If the use of a virtual co-working platform is not solely motivated by archiving intentions, but should support knowledge-integrating processes, regularly-documented results should be continuously subject to further processing and moderated evaluation in order to activate the bodies of knowledge and let them systematically flow into the project work.

The motivation on an individual level, a fundamental willingness to use the platform, and available capacities for the respective project participants determine whether and how continuously updated virtual infrastructures are used for knowledge-integrating collaboration in the project. Since separately set-up platforms can be completely new for all participants and thus go beyond the individual's customary information and work infrastructure, they require additional time and the willingness to learn how to benefit from the new structures and possibly discontinue or relearn personal work habits within the framework of the project.

4.2.2 Colloquia

Colloquia were held on both of the project's work levels. The small colloquia were focused on the involved research assistants and took place with the involvement of the project coordinators roughly every four weeks. All the project participants, meaning the research assistants and professors, were invited to the large quarterly colloquia.

Small Colloquia The format of the small colloquia initially had little influence on the sought synthesis of knowledge for a joint research result, but its integrative force strengthened over the course of the project. A greater exchange of knowledge could not take place solely by hearing the short oral reports and subsequent short discussions. In order to encourage this, expanded and largely free theory discussion was introduced, starting with text lectures. These discussion rounds that were focused on theory initially helped with understanding between the disciplines and sub-projects, but did not fulfil the playful-experimental interest in the project. They gave way to a practice-oriented exchange outside of the colloquia and in the form of studio or lab visits as well as *work in progress showcases* (cf. Sect. 4.2.4). Furthermore, the participants were able to engage in a more intensive exchange directly on their objects of research through *joint doing*, and thus understand their objects more deeply as a result.

The monthly colloquia also represented an instrument of coordination where the group could discuss organisational questions that arose from supporting the projects. Although these questions related to the process of actual research, usually involving formats for improving the integration performance, they were perceived more as an additional burden at the beginning. In the course of the project, the acceptance and joint responsibility for the co-shaping of the research design grew after the participants learned that they could influence the development of the project themselves in this way and thus also the results of the research. In the course of the project, the small colloquia established themselves as a framework for joint reflections on the research design on a meta-level (cf. Sect. 4.3).

Large Colloquia The large colloquia facilitated a transfer of knowledge between the research assistant level and the professors as well as between the sub-projects and the overall project level. The knowledge generated in the sub-projects between the respective professors and research assistants was largely inaccessible for the overall project level. In the large colloquia, it was possible to make everyone aware of the knowledge obtained in the sub-projects and to discuss it from new perspectives.

The presentations on the interim results of the sub-projects served as a basis for a more in-depth analysis in the small colloquia. Despite meetings for multiple hours, the available time was frequently not enough in order to conclude these discussions. Nonetheless, the additional perspectives of the professors from the other respective sub-projects initiated the integration of knowledge. These ideas were taken up and addressed collectively in other formats.

The effective involvement of the professors in the large colloquia required that the information flowed continuously from the level of the research assistants to the professor level within and between the partial projects and not only selectively on the occasion of the large colloquia.

A format for exchanging knowledge that was initiated by the research assistants consisted of the idea lectures held by the professors. They encouraged the integration of knowledge between the work levels. The subjects related more to the higher-level research question, for example, the prototyping methods anchored in the disciplines, ideal-typical processes and models of prototyping processes or various concepts in the terms model/prototype. The artefacts in *work in progress showcases*, for example, typical prototypes for the disciplines or specific interim results of sub-projects, served as *boundary objects*[10] for the encouragement of differentiation processes and synthesis.

Exchanging bodies of knowledge in the large colloquia between all the levels of research had a positive impact on the synthesis of knowledge for answering the joint research question. It can be assumed that this circulation of knowledge including all the participants also allowed new discoveries to be transferred to the individual, participating institutes and disciplines in terms of the two-directional impact of transdisciplinary knowledge integration.

4.2.3 Project Meetings with External Experts

On certain occasions, external prototyping experts[11] were invited to the large colloquia in order to analyse the developments in the sub-projects, comment on the interim results and enrich them with new points of view. They were supposed to provide inspiring perspectives on the subject or encourage creativity in the group as "free radicals". At the project meetings with guests, three formats were tried and

[10]*Boundary objects* in the literature on transdisciplinary research are central terms, concepts, ideas, plans, goals or also objects that are very important for all participants in regard to the collective research question or the collective issue, but are interpreted and understood differently. Their relevance as judged by all the participants establishes the interconnecting basis for communicating and mutually understanding the different meanings and interpretations of the *boundary objects*, whereby commonalities and differences arise from this. *Boundary objects* can initiate and promote the integration of knowledge. They act as transmitters and can, on the basis of differentiation and synthesis, lead to a collective understanding of the object itself, which represents a major basis for collaboration in transdisciplinary groups (cf. Bergmann et al. 2005, 44, among others). The term *boundary objects* was introduced by Star and Griesemer (1989).

[11]Cf. list of participating experts in this volume.

revealed varying integration potential: (1) theoretical presentations, (2) participation in the work in progress showcases of the research group and discussions on the basis of the presented objects, (3) workshops.

The discussions connected to the theoretical presentations given by the external lecturers were viewed as enriching. Additional contact to the lecturers probably would have caused them to have an even more extensive impact on the project. The frequency of these productive collisions in various stages of the project can be viewed retrospectively as potentially effective in order to accompany the project work not only selectively over the course of the entire project, but also continuously through an exchange with external lecturers as a source of inspiration.

As soon as *boundary objects* were available as a basis for discussion (format 2), the exchange between external lecturers and project participants intensified. As in the prototyping processes themselves, they made it possible to illustrate concepts, localise different views or problems and identify new ideas.

In the course of the workshops (format 3), the collaboration and exchange was the most intensive. This is how it was, for example, within the framework of a focus group[12] where two fundamental questions on prototyping were answered from the perspective of the practising engineers, designers, psychologists, humanities professors and philosophers. The project participants collected the perspectives gained in this multidisciplinary set-up, which amounted to an encyclopaedic integration, in a subsequent, project-internal workshop for analysis on the level of the intrapersonal integration of knowledge. The knowledge obtained together was also integrated into a joint text (cf. Israel et al. in this volume).

4.2.4 Workshop Visits and Public Showcases

Based on the participants' experiences of being able to discuss their research intensively and effectively by going into greater depth through joint design processes or visual demonstrations on objects, i.e. through *joint doing*, formats were introduced that encourage this type of transfer. Since the research in the sub-projects, with the exception of the mutual workshop visits mentioned above, was mostly conducted in a modular way, meaning in physically separate locations, the interim results were supposed to be combined in a general preview at regular intervals. This was achieved through formats at the interface to publicity such as elaborate presentations in public *work in progress showcases*. These exhibitions took place once during the "Long Night of the Sciences" in Berlin and twice at "Hybrid Talks", an independent format connected to the "Hybrid Plattform".[13] They presented a cross-section of the current state of the research. The integration

[12]Cf. Israel et al. on this method in this volume. The workshop took four hours and raised two questions: "What is prototyping?" and "What will the future of prototyping look like?".

[13]"Hybrid Talks" illuminate a subject in short presentations of roughly ten minutes from the perspective of multiple disciplines. The free exchange with the speakers takes place after the presentation.

potential of these public events was especially high since it led to concrete discussions on the exhibition objects of the sub-projects during the planning, development, execution and follow-up treatment. Another advantage of a joint exhibition room in which the objects are presented in parallel was seen in the juxtaposing that allows for direct comparison.

In the exchange with the interested public and thus with expertise outside of the field, the actual basis of research across disciplines is expanded even further to include disciplinary polyphony. "Hybrid Talks", for example, offered a framework for exchange with the Berlin creative economy and with external scientists and designers. The reflection on new and additional perspectives expanded the horizon of the participants' knowledge in regard to the joint research questions. Speaking about individual research in a public context helped with the finding of understandable vocabulary for field-specific and transdisciplinary results, which promoted the communication and language within the project team and thus the integration of knowledge. The "communicating of the scientific results in everyday language" (Krainer and Smetschka 2014, 78) is considered to be a central transdisciplinary competency that, within the framework of these events, advanced the multi-field competency of the scientists involved and thus also their ability to engage in intrapersonal integration.

4.2.5 Review Conferences

The review conferences were not planned in the application concept and were therefore not part of the project plan initially. The idea arose in the organisational and content constitution phase of the project. After the official commitment to fund the project and its start, the organisational structure had to be stabilised in a constitution phase and adjusted to the actual composition of the participants.[14] It was necessary to balance out the group dynamics and develop a joint scientific understanding of the project idea. These processes required intensive reflective support. This should allow that all the participants are aware of their own degree of freedom and that of others in the collaboration with respect to the described challenges in the scientific system (cf. Sect. 3). Based on this, a consensus should be reached on the

[14]The project and organisation structure should not be firmly set at the beginning of transdisciplinary research projects at universities. In this regard, it is necessary to briefly mention a general challenge for the organization of university transdisciplinary research projects: Since multiple months often pass between the filing of an application and the uncertain approval of an extensive project, the structures and participants for the project end up being not available at the time of the funding commitment. A basic structure for a transdisciplinary project must frequently be set up initially at universities, e.g. by hiring new staff, because the members of the application group do not have available capacities, for instance, or additional staff are required to realise the research plans. These upstream processes are both time- and resource-intensive and must largely be carried out in self-organisation and in a relatively short time before the official beginning of the project since in this phase the coordination (if planned) has usually not been determined yet.

achievable goals, and the research question formulated in the application for the project should be adjusted within the actual project group.

The colloquia for a certain period of time and at intervals of multiple weeks could not offer the framework for these processes. In the pertinent literature on transdisciplinary methods, reference is made to the time intensity and the additional effort for this (cf. Schmithals et al. 2011, 60, 70; Oppen and Müller 2014, 41). Furthermore, there was initially no awareness of this type of project constitution among the participants. Only during the project did the professor level inspire a change in course to self-reflection, which was supposed to provide more space and not be solely oriented on the contents. Particularly in the exchange with the accompanying researcher, a meta-level was established in the project where the project participants worked out important findings with regard to the collaboration and developed the idea of the review conferences with the coordinators. At intervals of roughly one year, two conferences took place, whereby the first, among other, important elements in the constitution phase were reviewed. Both conferences helped to ensure more in-depth understanding on each side, the professional-thematic exchange for the processing of the higher-level research question and the conceptual work on the joint ideas. Since it was the explicit wish of the research assistants[15] to dedicate themselves to the team building and intensive content work in a context without influential factors in everyday life, the two-day conferences took place in seminar rooms far away from the customary workplace. The isolation offered positive distance to the usual technologies and routine work methods and made the participants more open to new perspectives. A social aspect of the conferences that is important for group dynamics was also the fact that the evenings could be designed informally (cf. Schmithals et al. 2011, 32).

First Conference The first review conference was prepared and held in close collaboration with one external moderator and had two focal points: the optimisation of the project situation (team building, awareness of challenges in the context, cf. Sect. 3) and the content work on the joint research question. In the exercises and talks on the project situation, the participants addressed their personal scientific and creative backgrounds, interests, focal points in research and motivation. A stakeholder analysis revealed numerous influential factors through the large number of participants and the embedding of the project in various institutions. A potential analysis of the individual disciplines expanded the disciplinary characteristics of the cultures in the disciplines, raised awareness of stereotypical preconceptions and demonstrated specific strengths and weaknesses as well as supplementary and synergy potential between the disciplines. A capacity analysis

[15]The review conferences took place without the participation of the professors. This is to be understood against the backdrop that the research assistants wanted to meet far away from the influential factors in their daily life (cf. Sect. 3). The professor presence would have brought institutional connections with the implication of certain constraints and hierarchies into the context of a review conference. Only one guest professor from the UdK Berlin took part in the conferences since his function was to combine the two work levels.

made the available resources transparent. Starting from this, a new joint definition of the goal could be formulated on a minimum-maximum scale for the overall project level. The minimum goal was a traditional anthology of the results of the partial projects (encyclopaedic integration). The maximum goal included innovative sub-project results and knowledge synthesis in the form of a joint definition of prototyping, which should be published in an experimental format (intrapersonal integration). This self-formulated objective shows that the participants internalised the difference between multidisciplinarity and transdisciplinarity through the reflective process of the conference and derived clear transdisciplinary goals for the project. This first focal point of the conference also defined the project participants' awareness of the importance of continuous reflection on the work processes in the project over the long term, which had a positive impact on the ongoing collaboration and integration of knowledge.

The second focal point of the conference was on work related to the object of research. The idea of personal research interest brought new ideas for subjects and cooperation. The specific content work in regard to the joint issue of what prototyping is and how its concept can be reinterpreted was handled in a prototyping workshop: Initially, all the participants explained a typical prototyping process in their discipline and demonstrated this on the basis of the prototypes they brought with them. They also each presented a discipline-specific task, and in small discipline-mixed groups developed solutions in prototyping processes. Two observers documented the individual processes and simultaneously took down the central terms used in the communication. The respective scenario was recorded with a camera installed above the worktable.

The workshop was based on the concept of the *boundary object*,[16] which was the prototyping process in the conference. Individual prototyping processes formed the interconnecting basis for communicating and mutually understanding the different meanings and interpretations of *prototyping*, whereby commonalities and differences arise from this. In particular through the joint practical interaction, the diversity of the inherent concepts in the system of prototyping could be understood. The mutual understanding of each of the different meanings and processes of prototyping did not require any verbalisation. But it allowed in turn that previously, only implicitly available knowledge of each individual prototyping process and concept became known and explicit through joint experience in practice and was therefore to be verbalised for future collaboration.

The experiences and observations from these joint experiences in practice were evaluated after the end of the exercises: Terms such as prototype/model were considered in a differentiated way, and prototyping as a process rather than prototypes as objects moved to the centre of the analysis and consideration. A recapitulation and evaluation of the results from the first review conference took place in the second conference and were published in a joint article (cf. Exner et al. 2015).

---

[16]Cf. definition in Sect. 4.2.3, Footnote 10.

Second Conference The second review conference was designed analogously to the first. The only difference was that the conference was no longer co-designed and co-directed by an external moderator. We considered this to be a positive result of the intensive analysis on the meta-level for the project design. On its own, the research group implemented a high level of knowledge-integrating research work in collaboration with the coordinators after two years of joint research. In the second conference, the thematic focus was on the development of the concept of the final publication. Besides this publication, there is also a "package" that contains artefacts and multimedia elements that also offer additional access to research results on a popular-scientific level (cf. Ängeslevä et al. The Results of Rethinking Prototyping in this volume). The knowledge-integrating moment of the second conference was in the task of developing a joint concept for the final publication that must fundamentally be viewed as a significant element in transdisciplinary knowledge integration.

4.2.6 Joint Publications

Joint publishing of transdisciplinary research represents an important knowledge-integrating function that attaches significant relevance to transdisciplinary publishing.

In collective volumes of multidisciplinary projects, the integration of the knowledge and the synthesis are frequently left to the reader. Such a reader completes an intrapersonal integration through the lectures of individual, additively joined contributions and builds up cross references between texts that may be implicitly included, but were not addressed explicitly by the authors themselves or the editors. In the process of transdisciplinary publishing, this integration of knowledge does not take place outside of the research, but rather is done in the research process and also completed in the compositional and developmental process of the publication. The advantage is that the processes of intrapersonal integration primarily occurring in the individual also become visible and expressed as results in the text—and this finally makes the process of knowledge integration complete.

Making the results of the jointly completed synthesis of knowledge explicit in a publication requires that the project participants are willing to present experimental, i.e. atypical disciplinary solutions. This is because transdisciplinarily-formulated contents cannot be easily integrated into discipline-specific publication formats.

The "Rethinking Prototyping" project pursued a prototype for a transdisciplinary publication format in which the joint research work is visibly published for a broad public and for the corresponding scientific *communities*. Based on our experience, the value of such a transdisciplinary publication for the level of knowledge integration lies in three important functions.

Firstly, the final publication on the overall project level focuses the activities and the attention of the project participants on a joint project goal early on. It bundles and focuses the collaboration of the actors on a content and conceptual level and thus opens up a framework for action in which knowledge integration is a fundamental requirement for the successful achievement of this project goal—a simple anthology should not be produced at the end.

Secondly, the goal of developing a transdisciplinary publication format within the project team represents a joint task that opens up additional space for interaction and action in which the project participants enter into the exchange and into a process of collective creation. As a result of the fact that there is (still) no established transdisciplinary publication format, it was necessary to develop an appropriate prototype for the group. Since prototyping processes themselves achieve a high degree of knowledge integration by means of joint understanding through doing and intensive communication, high knowledge-integrating force can be attributed to the development of the publication format.

Thirdly, the joint writing must be viewed as an important cognitive means of intrapersonal knowledge integration: Through the process of writing, the bodies of knowledge that must be integrated on the text level are thought through again in more depth and renegotiated in terms of the goal of joint text production. The joint composition of texts on the higher-level research question is a method in order to re-express integrated knowledge on the intrapersonal level and to integrate all the discoveries of the involved authors in a semantic unit.

4.3 Accompanying Research

For successful collaboration in the transdisciplinary project group, it was a significant advantage that "Rethinking Prototyping" was analysed in the framework of the accompanying research on the "Hybrid Plattform" by the sociologist Maria Oppen (Social Science Research Center Berlin, WZB) in a dialogic form. The focus of her research was particularly on the communication processes in connection with the existing project structures.

The interviews and responses of the scientists offered a bird's eye view of the work processes in the project that led to a critical and productive self-evaluation. The accompanying study exposed very specific problems, bottlenecks as well as opportunities and potential. Of particular importance for the accompanying research in relation to the research project was:

> The respective abstract concept of interdisciplinarity considered to be self-explanatory [...] was deconstructed and filled with specific building blocks of action by using observations from accompanying research. (Oppen and Müller 2014, 57f.)

Accordingly, it was possible to have a positive impact on the identified problems in the ongoing course of the project. Consideration of the processes from a self-reflective perspective, which were revealed to the scientists in individual talks with the researcher, created an awareness of the challenges and problems in the scientist's own transdisciplinary research process, allowing and promoting active co-designing of the collaboration. In turn, this increased identification with the project and the acceptance of the developed and applied formats.

The process of self-clarification (cf. Heintel 2006 is a pre-requisite for the success of transdisciplinary research processes (cf. Lerchster and Lesjak 2014, 82). The connected "ability to self-analyse" (Lerchster and Lesjak 2014, 82) also falls under the multi-field competency described by Ropohl, which promotes intrapersonal integration. Establishing a corresponding functional reflection space as a meta-level within a research group may be the task of a (if possible) professionally trained intermediary or moderator (cf. Oppen and Müller 2014, 45). This person should have an awareness of the special challenges in transdisciplinary research as discussed, for example, in the accompanying research and here in Sect. 3, and be sensitised for the socio-communicative dynamism in the group (cf. Oppen and Müller 2014, 45). It would be an advantage if the person has "the ability 'to think outside' the traditional disciplinary cultures" and is "familiar with diverging worldviews and conditions for producing knowledge" (Oppen and Müller 2014, 45) in order to also reflect on these individually in the process and be able to mirror the project participants. In some projects, there are researchers who can adopt this role in part or in full. This task was partially handled by the coordinators and the accompanying research in this project. In the course of the project, this function was increasingly supported by the project participants who learned the multi-field competency for transdisciplinary action and implemented this in the co-designing of the process.

4.4 *Conclusion: Factors in Successful Knowledge Integration*

Some of the discovered factors that are beneficial for knowledge integration can apply to other transdisciplinary research projects in the university context. A normative consolidation of our procedure would be misplaced, however, since each transdisciplinary project is designed differently, has its "own logic and dynamics" (Oppen and Müller 2014, 65) and requires a certain flexibility in its execution and various approaches. This was also seen in the finding, testing, modification, iterative repetition and occasional problems in the methods and formats implemented in "Rethinking Prototyping". In this sense, each case of transdisciplinary research can itself be considered to be a prototyping process. Transdisciplinary research is a process of continuous optimisation, and after the conclusion of the project one can learn from it as a prototype for future processes. Retrospective reflection on the most important formats in collaboration makes it possible to determine the following factors that were required in particular for the integration of knowledge in the course of this special project.

Meta-Discourse: Reflections on the Research Process The shortcomings in the lack of methodology for interdisciplinary projects as quoted from Ropohl at the beginning (cf. Ropohl 2010, 5) could be balanced out in this project through continuous discourse accompanying the project on the meta-level, since the research process itself became the object of reflection. As the prototyping processes are reflective, communicative, iterative and recursive processes, this reflection, inherent in prototyping, is also mirrored in all transdisciplinary research in our opinion. It serves to optimise the research process, the research design and, in some circumstances, even the research question.

This level of self-reflection led to an improvement in the project participants' cooperative actions in this project. In the course of the three-year reflective process, they tested transdisciplinary, as opposed to multidisciplinary, work and co-designed the corresponding research design themselves. In particular, the dialogic accompanying research for the project reflected a significant role in the formation of this reflective meta-level in the project. Last but not least, discussions with the accompanying researcher promoted the development of a productive communication and cooperation culture.

Such meta-reflection requires a high degree of self-awareness from the project participants since they themselves are also involved in the process of collaboration that they should reflect upon and co-design. This means that the participants must be aware of the specific challenges in the overall context of the project in order to be able to judge, for example, their degree of freedom to design the process and the capacities.

The co-determination and freedom to design that result from this meta-reflection help to build an identity and strengthen the collective awareness for collaboration on the joint research question. As a result, the participants' willingness to concentrate on the transdisciplinary object of research at the edge of their discipline grows. The discourse on the meta-level also encourages the formation of integrative competencies among the participants in terms of the designed integration of knowledge and synthesis. In summary, the discursive meta-level in regard to discipline-overarching collaboration can be viewed as one of the central constituent factors in transdisciplinary research with a high degree of integration force.

Flexible Question and Openness for Results In the process of the spirally-running integration of knowledge (cf. Sect. 2.2), the research question is repeatedly scrutinised and modified, which can lead to a re-orientation in the project-related research. The shifting of the research focal point on the overall level from the theoretical approach to the gaining of knowledge through joint practical doing, as in "Rethinking Prototyping", provided an example of this modification. If the question or the objective is shifted in the course of the project, it is important that this change is prepared as collectively as possible in the project team and accepted by the largest possible number of project participants. If the personnel in the application group for a research project differ from the actual team for the research project, such a change can also encourage greater identification with the project and increase the motivation to work on a joint question. The research interests of the individual participants may also shift during the course of the project, which can likewise require an expansion or adjustment of the developed research question in order to maintain the greatest possible intersection of the overall research interest. This means that the project participants must exhibit a certain degree of flexibility in their handling of the transdisciplinary research question and a fundamental openness for the results in the project.

Prototyping as a Method and "Understanding by Doing" The shift in the focal point of the work from the theoretical level to the practical level brought about an adjustment in the purely verbal exchange of non-verbal elements in the *joint doing*. In this connection, the joint prototyping proved to be a central knowledge-integrating method between the disciplines and thus a general method in transdisciplinary collaboration.

Implicit knowledge can be revealed by joint prototyping without verbal concepts since something becomes understandable, objectified and comprehensible through prototyping for which initially there is no joint language per se, as is typically the rule in multidisciplinary contexts. The method of prototyping in the work with boundary objects as, for example, in the workshop from the first review conference (cf. Sect. 4.2.5) can make mutual understanding easier and shorten the length of the formation of a linguistic basis of understanding. In this way, prototyping initially renders linguistic translation superfluous. The process of joint prototyping also leads to the development of prototypes that can represent a partial solution for a problem posed within the context of transdisciplinary research and represents an important basis for the general discursive integration of knowledge since:

> [f]irstly, the sensory-specific, motor-related, interactive reference to physical objects makes it possible for actors to create, combine, destroy and discard mental models of meaning units or speak about them and reflect on them (Adenauer and Petruschat 2012, 17).

Besides the previously described function in regard to the meta-discourse on transdisciplinary collaboration, this possibility of using prototyping underscores its knowledge-integrating potential once again.

Joint Spaces An important factor for the intensity of knowledge integration is space. Space is understood, on the one hand, as jointly defined conceptual mental and reflective spaces and, on the other, as real spaces in which scientists and creators act. Conceptual spaces fundamentally take shape when actors who have participated in mental processes are not at the same location. In the project, however, it was difficult to fill these conceptual spaces with life during the phases of distance. Since the knowledge integration is completed in these conceptual spaces, it was necessary to bring the participants together at collective locations that intensified the cognitive processes through personal exchange and joint doing. The conception and realisation of exhibitions, workshops and teaching offers, for example, satisfied this need.

The greatest intensity in the joint work was achieved in situations in which the group met at a secluded location shielded from systematically conditioned influential factors for a longer period of time. This made it possible for the participants to mentally enter the conceptual room without any mental disruption and spend time there. From this it is possible to derive that for transdisciplinary research at least one collective space is of great significance for the research group in order to maintain the collective conceptual space.

This space should have important properties that distinguish it from the project participants' normal workplaces. Suitable spaces in this sense are "exterritorial [...] spaces" (Oppen and Müller 2014, 46), which are not defined by one specific discipline and its research and working habits. In these spaces the attempt is made to relativise the existing hierarchies and the everyday work does not interfere with concentration on the transdisciplinary research (cf. Oppen and Müller 2014, 46). Ideally, these spaces allow for collaboration based on interaction and communication as well as withdrawn, concentrated work since this dualism is essential for creative and innovative stimulus in transdisciplinary work (cf. Phillips 2014, 99). The more time that is spent at these locations in order to open up collective knowledge space, the higher the degree of knowledge integration. Such space could be defined by a very independent work culture and create a truly transdisciplinary, third space between the participating disciplines. The Hybrid Lab,[17] which was available for the project in the last third of the project period, offered such a space. The review conferences corresponded most of all to this ideal space where at all times a conceptual or practical task was handled collectively and the space constellation isolated from the usual work environment brought about intensive intrapersonal integration. This had an impact beyond the conference itself, extending to the individual workspaces of the project participants, since an unparalleled rise in the capacities of the group was observed in the initial weeks after the review conference.

Besides these separate real spaces within the project, spaces at the interface to the public can also develop conceptual space and thus promote the integration of knowledge. There are spaces like the created *showcases* and exhibitions that expand the communication with external perspectives and focus and promote exchange with outsiders. The communication processes here, which were foreign to the disciplines and outside of science, require an expansion of the individual language on the object of research and promote the verbalisation and exchange of the gained experiences—central aspects of transdisciplinary multi-field competency.

[17]The Hybrid Lab is a space within the "Hybrid Plattform", which places this at the disposal of transdisciplinary project groups, among others. Various project partners and promoters, scientists and artistic staff at the UdK Berlin and the TU Berlin, members of the "Hybrid Plattform" Association (Hybrid Plattform e.V.) and the public come together here for the joint work or events. The Hybrid Lab is located on the Charlottenburg campus in the building EB of the TU Berlin.

5 Outlook

Our work demonstrated, in addition to other factors, the central importance of reflective analysis in transdisciplinary research on the micro level. Concluding this third part of the book, we also want to highlight its significance on the macro level and understand our reflections in a larger context.

The meta-reflective level taken up in this text on the project counters a phenomenon that is described as black boxing in the literature. At the end of an intensive research project, the sum of the steps taken, the entire way to the goal, appears to be self-explanatory (cf. Bammé and Spök 2014, 42): "[T]he process in the course of which the consensus was jointly produced is increasingly forgotten. It is invisible, so to say" (Bammé and Spök 2014, 42). We are persuaded that the interactions, dead ends and partial failure of this development process in successful transdisciplinary research should be reflected upon and documented for three reasons: (1) the reflection makes us aware of unconscious processes and contributes to the participants' reinforcement of acquired multi-field competencies. The documentation sets the findings in the reflective processes and develops a knowledge archive of experiences that all participants can rely on in future projects. (2) The documentation is also an orientation aid for future coordinators and supporters of transdisciplinary projects. (3) If each completed project is understood as a prototype (and simultaneously a product) of transdisciplinary research, this encourages a global prototyping process in transdisciplinary research in which the realisation strategies are tested, evaluated and optimised. In this context of prototyping, we can confirm the thoughts of Hubert Laitko:. He argues that trans- or "interdisciplinarity is not a local quality of the individual research process, but rather a global holistic disposition in an entire scientific system that is produced and reproduced by this" (Laitko 2011, 8). This global disposition must ensure that an individual's ability to think transdisciplinarily is formed systematically, continuously and in a controlled way. As a result, the goal is to enable participants in the transdisciplinary processes to create a synthesis, on the one hand, and to increase their potential, on the other, by bringing new and stimulating knowledge into the individual disciplines via intrapersonal integration. Furthermore, the prototyping process of transdisciplinary research must be theoretically emphasised and supported by the sub-systems of the scientific system.

For the "Rethinking Prototyping" project, the close collaboration with the "Hybrid Plattform" represented a supporting systemic requirement that is rarely found in the university context. Finally, reference is made to the particularly advantageous situation of coordination for the design of the meta level in the "Rethinking Prototyping" project, which emphasises the model-like character of the project. The success of the project was not solely placed in the hands of the scientists and creators; coordination was planned from the beginning. This was set up on the transdisciplinary "Hybrid Plattform" of the TU Berlin and the UdK Berlin, which facilitated the reflecting-moderating support of the project. Both the platform and the project benefited from the synergy effects that resulted from the spatial proximity, organisational interweaving and the regular exchange of content. Particularly valuable for the coordinators was the access to the experiences of other transdisciplinary projects and participation in the accompanying research of the "Hybrid Plattform". Retrospectively, we view the "Hybrid Plattform" as making an important contribution to the global disposition of transdisciplinarity in the scientific system as postulated by Laitko.

In conclusion, it should be noted that: This project did not by any means run on its own, but was also not left alone in order to develop the desired transdisciplinary added value. The will and motivation that most of the project participants demonstrated in this project was an important basis for successfully conducting it. The additionally developed knowledge-integrating formats and instruments, the particular spatial advantages, the flexibly managed overall interest in the research and the significant meta level of the process reflection prevented this project from becoming an ambivalent interdisciplinary experience. Rather, it is possible to say here in summary that this complex and diverse project led to a successful transdisciplinary conclusion of the project.

Acknowledgments We thank our reviewers Johann Habakuk Israel and Julia Warmers for the time and effort that they invested into the review of our manuscript, and for their helpful comments and suggestions. Special thanks also go to Julia Klauer who created the figures for this text.

References

Adenauer, J., & Petruschat, J. (Eds.). (2012). *Prototype! physical, virtual, hybrid, smart. Tackling new challenges in design and engineering*. form+zweck, Berlin.

Arnold, M., Gube, V., & Wieser, B. (2014). Interdisziplinär forschen. In G. Dressel, W. Berger, K. Heimerl, & V. Winiwarter (Eds.), *Interdisziplinär und transdisziplinär Forschen. Praktiken und Methoden* (pp. 105–119). Bielefeld: Transcript Verlag.

Balsiger, P. W. (2005). *Transdisziplinarität. Systematisch-vergleichende Untersuchung disziplinübergreifender Wissenschaftspraxis*. Munich: Wilhelm Fink Verlag.

Bammé, A., & Spök, A. (2014). Probleme wahrnehmen und strukturieren. In G. Dressel, W. Berger, K. Heimerl, & V. Winiwarter (Eds.), *Interdisziplinär und transdisziplinär Forschen. Praktiken und Methoden* (pp. 37–49). Bielefeld: Transcript Verlag.

Bergmann, M., Brohmann, B., Hoffmann, E., Loibl, C. M., Rehaag, R., Schramm, E., et al. (2005). *Qualitative criteria of transdisciplinary research. A guide for the formative evaluation of research projects*, ISOE-Studientexte, No. 13/English version. Institute for Social-Ecological Research (ISOE) GmbH, Frankfurt am Main.

Dressel, G., Heimerl, K., Berger, W., & Winiwarter, V. (2014). Interdisziplinäres und transdisziplinäres Forschen organisieren. In G. Dressel, W. Berger, K. Heimerl, & V. Winiwarter (Eds.), *Interdisziplinär und transdisziplinär Forschen. Praktiken und Methoden* (pp. 207–2012). Bielefeld: Transcript Verlag.

Exner, K., Änglesleva, J., Bähr, B., Nagy, E., Lindow, K., & Stark, R. (2015). A transdisciplinary perspective on prototyping. In *International Conference on Engineering, Technology and Innovation* (Belfast, Ireland). Institute of Electrical and Electronics Engineers. (in press).

Heintel, P. (2006). Über drei Paradoxien der T-Gruppe. In Heintel, P. (Ed.),*Betrifft TEAM Dynamische Prozesse in Gruppen* (pp. 196–243). Wiesbaden: VS Verlag für Sozialwissenschaften.

Krainer, L., & Smetschka, B. (2014). Ein Forschungsteam finden. In Gert Dressel, Wilhelm Berger, Katharina Heimerl, & Verena Winiwarter (Eds.), *Interdisziplinär und transdisziplinär Forschen. Praktiken und Methoden* (pp. 65–78). Bielefeld: Transcript Verlag.

Laitko, H. (2011). Interdisziplinarität als Thema der Wissenschaftsforschung. In *LIFIS ONLINE* (26. 10. 2011). www.leibniz-institut.de. ISSN 1864-6972.

Lerchster, R., & Lesjak, B. (2014). Forschungsteams organisieren. Eine gruppendynamische Perspektive. In G. Dressel, W. Berger, K. Heimerl & V. Winiwarter (Eds.) *Interdisziplinär und transdisziplinär Forschen. Praktiken und Methoden* (pp. 79–90). Bielefeld: Transcript Verlag.

Mittelstraß, J. (2003). *Transdisziplinarität-Wissenschaftliche Zukunft und institutionelle Wirklichkeit*. Constance: UVK Universitätsverlag Konstanz GmbH.

Mittelstraß, J. (2008). Wenn sich die Forschung bewegt... Über die Universität und die Notwendigkeit einer Reform unseres Wissenschaftssystems. In *LIFIS ONLINE* (08. 12. 2008). www.leibniz-institut.de. ISSN 1864-6972.

Oppen, M., & Müller, C. (2014). *Von der Kollision zur Kooperation. Zusammenarbeit zwischen künstlerisch-gestaltenden und technisch-naturwissenschaftlichen Disziplinen*. epubli, Berlin. Free pdf-download: http://www.hybrid-plattform.org/images/Von_der_Kollision_zur_Kooperation.pdf.

Phillips, M. N. (2014). Interdisziplinarität als Vehikel für Kreativität und Innovation. In C. Schier & E. Schwinger (Eds.), *Interdisziplinarität und Transdisziplinarität als Herausforderung akademischer Bildung. Innovative Konzepte für die Lehre an Hochschulen und Universitäten* (pp. 95–103). Bielefeld: Transcript Verlag.

Ropohl, G. (2010). Jenseits der Disziplinen–Transdisziplinarität als neues Paradigma. In *LIFIS ONLINE* (21. 03. 2010). www.leibniz-institut.de. ISSN 1864-6972.

Schmithals, J., Loibl, C., Dienel, H.-L., & von Braun, C.-F. (2011). Kleines Einmaleins inter- und transdisziplinärer Forschungskooperation. Anspruch und Wirklichkeit in der Kooperation zwischen Wissenschaft und Praxis. Empirische Befunde und Handlungsempfehlungen. in Andrea von Braun Stiftung ed. *Briefe zur Interdisziplinarität*. Munich: Oekom Verlag.

Star, S. L., & Griesemer, J. (1989). Institutional ecology, 'translations' and boundary objects amateurs and professionals in Berkeley's Museum of Vertebrate Zoology, 1907–39. In *Social studies of science* (Vol. 19(3), pp. 387–420). New York: Sage Publications.

Stark, C. (2009). Funktionalismus. In G. Kneer & M. Schroer (Eds.), *Handbuch Soziologische Theorien* (pp. 161–178). Wiesbaden: VS Verlag für Sozialwissenschaften.